Yakov Perelman's
Physics For Entertainment.
By
Yakov Perelman

Physics for Entertainment

By Yakov Perelman

Published in 1913, a best-seller in the 1930s and long out of print, *Physics for Entertainment* was translated from Russian into many languages and influenced science students around the world. Among them was Grigori Yakovlevich Perelman, the Russian mathematician (unrelated to the author), who solved the Poincaré conjecture, and who was awarded and rejected the Fields Medal. Grigori's father, an electrical engineer, gave him Physics for Entertainment to encourage his son's interest in mathematics. In the foreword, the book's author describes the contents as "conundrums, brain-teasers, entertaining anecdotes, and unexpected comparisons," adding, "I have quoted extensively from Jules Verne, H. G. Wells, Mark Twain and other writers, because, besides providing entertainment, the fantastic experiments these writers describe may well serve as instructive illustrations at physics classes." The book's topics included how to jump from a moving car, and why, "according to the law of buoyancy, we would never drown in the Dead Sea." Ideas from this book are still used by science teachers today. Yakov Isidorovich Perelman died in the siege of Leningrad in 1942.

CONTENTS

Chapter Three

ATMOSPHERIC RESISTANCE

Chapter Four

ROTATION. "PERPETUAL MOTION" MACHINES

Chapter Five

PROPERTIES OF LIQUIDS AND GASES

Chapter Six

HEAT

Chapter Seven

LIGHT

Chapter Eight

REFLECTION AND REFRACTION

Chapter Nine

VISION

Chapter Ten

SOUND AND HEARING

FROM THE AUTHOR'S FOREWORD
TO THE 13th EDITION

The aim of this book is not so much to give you some fresh knowledge, as to help you "learn what you already know". In other words, my idea is to brush up and liven your basic knowledge of physics, and to teach you how to apply it in various ways. To achieve this purpose conundrums, brain-teasers, entertaining anecdotes and stories, amusing experiments, paradoxes and unexpected comparisons—all dealing with physics and based on our everyday world and sci-fic—are afforded. Believing sci-fic most appropriate in a book of this kind, I have quoted extensively from Jules Verne, H. G. Wells, Mark Twain and other writers, because, besides providing entertainment, the fantastic experiments these writers describe may well serve as instructive illustrations at physics classes.

I have tried my best both to arouse interest and to amuse, as I believe that the greater the interest one shows, the closer the heed one pays and the easier it is to grasp the meaning—thus making for better knowledge.

However, I have dared to defy the customary methods employed in writing books of this nature. Hence, you will find very little in the way of parlour tricks or spectacular experiments. My purpose is different, being mainly to make you think along scientific lines from the angle of physics, and amass associations with the variety of things from everyday life. I have tried in rewriting the original copy to follow the principle that was formulated by Lenin thus: "The popular writer leads his reader towards profound thoughts, towards profound study, proceeding from simple and generally known facts; with the aid of simple argu-

9

ments or striking examples he shows the main *conclusions* to be drawn from those facts and arouses in the mind of the thinking reader ever newer questions. The popular writer does not presuppose a reader that does not think, that cannot or does not wish to think; on the contrary, he assumes in the undeveloped reader a serious intention to use his head and *aids* him in his serious and difficult work, leads him, helps him over his first steps, and *teaches* him to go forward independently. (*Collected Works*, Vol. 5, p. 311, Moscow 1961.)

Since so much interest has been shown in the history of this book, let me give you a few salient points of its "biography".

Physics for Entertainment first appeared a quarter of a century ago, being the author's first-born in his present large family of several score of such books. So far, this book—which is in two parts—has been published in Russian in a total print of 200,000 copies. Considering that many are to be found on the shelves of public libraries, where each copy reaches dozens of readers, I daresay that millions have read it. I have received letters from readers in the furthermost corners of the Soviet Union.

A Ukrainian translation was published in 1925, and German and Yiddish translations in 1931. A condensed German translation was published in Germany. Excerpts from the book have been printed in French—in Switzerland and Belgium—and also in Hebrew—in Palestine.

Its popularity, which attests to the keen public interest displayed in physics, has obliged me to pay particular note to its standard, which explains the many changes and additions in reprints. In all the 25 years it has been in existence the book has undergone constant revision, its latest edition having barely half of the maiden copy and practically not a single illustration from the first edition.

Some have asked me to refrain from revision, not to be compelled "to buy the new revised edition for the sake of a dozen or so new pages". Scarcely can such considerations absolve me of my obligation constantly to improve this book in every way. After all *Physics for Entertainment* is not a work of fiction. It is a book on science—be it even popular science—and the subject taken, physics, is enriched even in its fundamentals with every day. This must necessarily be taken into consideration.

On the other hand, I have been reproached more than once for failing to deal in this book with questions such as the latest achievements in radio engineering, nuclear fission, modern theories and the like. This springs from a misunderstanding. This book has a definite purpose; it is the task of other books to deal with the points mentioned.

Physics for Entertainment has, besides its second part, some other associated books of mine. One, *Physics at Every Step*, is intended for the unprepared layman who has still not embarked upon a systematic study of physics. The other two are, on the contrary, for people who have gone through a secondary school course in physics. These are *Mechanics for Entertainment* and *Do You Know Your Physics?*, the last being the sequel, as it were, to this book.

1936   *Y. Perelman*

CHAPTER ONE

SPEED AND VELOCITY. COMPOSITION
OF MOTIONS

HOW FAST DO WE MOVE?

A good athlete can run 1.5 km in about 3 min 50 sec—the 1958 world record was 3 min 36.8 sec. Any ordinary person usually does, when walking, about 1.5 metres a second. Reducing the athlete's rate to a common denominator, we see that he covers seven metres every second. These speeds are not absolutely comparable though. Walking, you can keep on for hours on end at the rate of 5 km. p.h. But the runner will keep up his speed for only a short while. On quick march, infantry move at a speed which is but a third of the athlete's, doing 2 m/sec, or 7 odd km.p.h. But they can cover a much greater distance.

I daresay you would find it of interest to compare your normal walking pace with the "speed" of the proverbially slow snail or tortoise. The snail well lives up to its reputation, doing 1.5 mm/sec, or 5.4 metres p.h.—exactly one thousand times less than your rate. The other classically slow animal, the tortoise, is not very much faster, doing usually 70 metres p.h.

Nimble compared to the snail and the tortoise, you would find yourself greatly outraced when comparing your own motion with other motions—even not very fast ones—that we see all around us. True, you will easily outpace the current of most rivers in the plains and be a pretty good second to a moderate wind. But you will successfully vie with a fly, which does 5 m/sec, only if you don skis. You won't overtake a hare or a hunting dog even when riding a fast horse and you can rival the eagle only aboard a plane.

Still the machines man has invented make him second to none for speed. Some time ago a passenger hydrofoil ship, capable of 60-70 km. p.h., was launched in the U.S.S.R. (*Fig. 1*). On land you can move faster

Fig. 1. Fast passenger hydrofoil ship

than on water by riding trains or motor cars—which can do up to 200 km. p.h. and more (*Fig. 2*). Modern aircraft greatly exceed even these speeds. Many Soviet air routes are serviced by the large TU-104

Fig. 2. New Soviet ZIL-111 motor car

(*Fig. 3*) and TU-114 jet liners, which do about 800 km. p.h. It was not so long ago that aircraft designers sought to overcome the "sound barrier", to attain speeds faster than that of sound, which is 330 m/sec, or 1,200 km. p.h. Today this has been achieved. We have some small but very fast supersonic jet aircraft that can do as much as 2,000 km.p.h.

There are man-made vehicles that can work up still greater speeds. The initial launching speed of the first Soviet sputnik was about

Fig. 3. TU-104 jet airliner

8 km/sec. Later Soviet space rockets exceeded the so-called "escape" velocity, which is 11.2 km/sec at ground level.

The following table gives some interesting speed data.

A snail	1.5 mm/sec	or	5.4 metres p.h.
A tortoise	20 "	or	70 "
A fish	1 m/sec	or	3.5 km. p.h
A pedestrian	1.4 "	or	5 "
Cavalry, pacing	1.7 "	or	6 "
" trotting	3.5 "	or	12.6 "
A fly	5 "	or	18 "
A skier	5 "	or	18 "
Cavalry, galloping	8.5 "	or	30 "
A hydrofoil ship	16 "	or	58 "
A hare	18 "	or	65 "
An eagle	24 "	or	86 "
A hunting dog	25 "	or	90 "
A train	28 "	or	100 "
A ZIL-111 passenger car	50 "	or	170 "
A racing car (record)	174 "	or	633 "
A TU-104 jet airliner	220 "	or	800 "
Sound in air	330 "	or	1,200 "
Supersonic jet aircraft	550 "	or	2,000 "
The earth's orbital velocity	30,000 "	or	108,000 "

RACING AGAINST TIME

Could one leave Vladivostok by air at 8 a.m. and land in Moscow at 8 a.m. on the same day?

I'm not talking through my hat. We can really do that. The answer lies in the 9-hour difference in Vladivostok and Moscow zonal times. If our plane covers the distance between the two cities in these 9 hours, it will land in Moscow at the very same time at which it took off from Vladivostok. Considering that the distance is roughly 9,000 kilometres, we must fly at a speed of $9,000:9=1,000$ km. p.h., which is quite possible today.

To "outrace the Sun" (or rather the earth) in Arctic latitudes, one can go much more slowly. Above Novaya Zemlya, on the 77th parallel, a plane doing about 450 km. p.h. would cover as much as a definite point on the surface of the globe would cover in an identical space of time in the process of the earth's axial rotation. If you were flying in such a plane you would see the sun suspended in immobility. It would never set, provided, of course, that your plane was moving in the proper direction.

It is still easier to "outrace the Moon" in its revolution around the earth. It takes the moon 29 times longer to spin round the earth than it takes the earth to complete one rotation (we are comparing, naturally, the so-called "angular", and not linear, velocities). So any ordinary steamer making 15-18 knots could "outrace the Moon" even in the moderate latitudes.

Mark Twain mentions this in his *Innocents Abroad*. When sailing across the Atlantic, from New York to the Azores "... we had balmy summer weather, and nights that were even finer than the days. We had the phenomenon of a full moon located just in the same spot in the heavens at the same hour every night. The reason for this singular conduct on the part of the moon did not occur to us at first, but it did afterward when we reflected that we were gaining about twenty minutes every day, because we were going east so fast—we gained just enough every day to keep along with the moon."

THE THOUSANDTH OF A SECOND

For us humans, the thousandth of a second is nothing from the angle of time. Time intervals of this order have only started to crop up in some of our practical work. When people used to reckon the time according to the sun's position in the sky, or to the length of a shadow (*Fig. 4*), they paid no heed to minutes, considering them even unworthy

Fig. 4. How to reckon the time according to the position of the sun (left), and by the length of a shadow (right)

of measurement. The tenor of life in ancient times was so unhurried that the timepieces of the day—the sun-dials, sand-glasses and the like—had no special divisions for minutes (*Fig. 5*). The minute hand first appeared only in the early 18th century, while the second sweep came into use a mere 150 years ago.

But back to our thousandth of a second. What do you think could happen in this space of time? Very much, indeed! True, an ordinary train would cover only some 3 cm. But sound would already fly 33 cm and a plane half a metre. In its orbital movement around the sun, the earth would travel 30 metres. Light would cover the great distance of 300 km. The minute organisms around us wouldn't think the thousandth of a second so negligible an amount of time—if they could think of course. For insects it is quite a tangible interval. In the space of a second a mosquito flaps its wings 500 to 600 times. Consequently in the space of a thousandth of a second, it would manage either to raise its wings or lower them.

We can't move our limbs as fast as insects. The fastest thing we can do is to blink our eyelids. This takes place so quickly that we fail even to notice the transient obscurement of our field of vision. Few know, though, that this movement, "in the twinkling of an eye"—which has

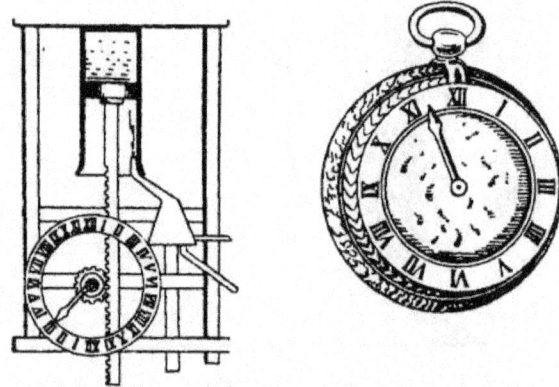

Fig. 5. An ancient water clock (left) and an old pocket-watch (right). Note that neither has the minute hand

become synonymous for incredible rapidity—is quite slow if measured in thousandths of a second. A full "twinkling of an eye" averages—as exact measurement has disclosed—two-fifths of a second, which gives us 400 thousandths of a second. This process can be divided into the following stages: firstly, the dropping of the eyelid which takes 75-90 thousandths of a second; secondly, the closed eyelid in a state of rest, which takes up 130-170 thousandths; and, thirdly, the raising of the eyelid, which takes about 170 thousandths.

As you see, this one "twinkling of an eye" is quite a considerable time interval, during which the eyelid even manages to take a rest. If we could photograph mentally impressions lasting the thousandth of a second, we would catch in the "twinkling of an eye" two smooth motions of the eyelid, separated by a period during which the eyelid would be at rest.

Generally speaking, the ability to do such a thing would completely transform the picture we get of the world around us and we would see the odd and curious things that H. G. Wells described in his *New Accelerator*. This story relates of a man who drank a queer mixture which caused him to see rapid motions as a series of separate static phenomena. Here are a few extracts.

"'Have you ever seen a curtain before a window fixed in that way before?'

"I followed his eyes, and there was the end of the curtain, frozen, as it were, corner high, in the act of flapping briskly in the breeze.

"'No,' said I, 'that's odd.'

"'And here,' he said, and opened the hand that held the glass. Naturally I winced, expecting the glass to smash. But so far from smashing it did not even seem to stir; it hung in mid-air—motionless. 'Roughly speaking,' said Gibberne, 'an object in these latitudes falls 16 feet in a second. This glass is falling 16 feet in a second now. Only you see, it hasn't been falling yet for the hundredth part of a second. [Note also that in the first hundredth of the first second of its downward flight a body, the glass in this case, covers not the hundredth part of the distance, but the 10,000th part (according to the formula $S = 1/2\ gt^2$). This is only 0.5 mm and in the first thousandth of the second it would be only 0.01 mm.]

"'That gives you some idea of the pace of my Accelerator.' And he waved his hand round and round, over and under the slowly sinking glass.

"Finally he took it by the bottom, pulled it down and placed it very carefully on the table. 'Eh?' he said to me, and laughed....

"I looked out of the window. An immovable cyclist, head down and with a frozen puff of dust behind his driving-wheel, scorched to overtake a galloping *char-à-banc* that did not stir....

"We went out by his gate into the road, and there we made a minute examination of the statuesque passing traffic. The top of the wheels and some of the legs of the horses of this *char-à-banc*, the end of the whip lash and the lower jaw of the conductor—who was just beginning to yawn—were perceptibly in motion, but all the rest of the lumbering conveyance seemed still. And quite noiseless except for a faint rattling that came from one man's throat! And as parts of this frozen edifice there were a driver, you know, and a conductor, and eleven people!...

"A purple-faced little gentleman was frozen in the midst of a violent struggle to refold his newspaper against the wind; there were many evidences that all these people in their sluggish way were exposed to a considerable breeze, a breeze that had no existence so far as our sensations went....

"All that I had said, and thought, and done since the stuff had begun to work in my veins had happened, so far as those people, so far as the world in general went, in the twinkling of an eye...."

Would you like to know the shortest stretch of time that scientists can measure today? Whereas at the beginning of this century it was only the 10,000th of a second, today the physicist can measure the 100,000 millionth of a second; this is about as many times less than a second as a second is less than 3,000 years!

THE SLOW-MOTION CAMERA

When H. G. Wells was writing his story, scarcely could he have ever thought he would see anything of the like. However he did live to see the pictures he had once imagined, thanks to what has been called the slow-motion camera. Instead of 24 shots a second—as ordinary motion-picture cameras do—this camera makes many times more. When a film shot in this way is projected onto the screen with the usual speed of 24 frames a second, you see things taking place much more slowly than normally—high jumps, for instance, seem unusually smooth. The more complex types of slow-motion cameras will almost simula H. G. Wells's world of fantasy.

WHEN WE MOVE ROUND THE SUN FASTER

Paris newspapers once carried an ad offering a cheap and pleasant way of travelling for the price of 25 centimes. Several simpletons mailed this sum. Each received a letter of the following content:

"Sir, rest at peace in bed and remember that the earth turns. At the 49th parallel—that of Paris—you travel more than 25,000 km a day. Should you want a nice view, draw your curtain aside and admire the starry sky."

The man who sent these letters was found and tried for fraud. The story goes that after quietly listening to the verdict and paying the fine demanded, the culprit struck a theatrical pose and solemnly declared, repeating Galileo's famous words: "It turns."

He was right, to some extent, after all, every inhabitant of the globe "travels" not only as the earth rotates. He is transported with still greater speed as the earth revolves around the sun. *Every second* this planet of ours, with us and everything else on it, moves 30 km in space, turning meanwhile on its axis. And thereby hangs a question not devoid of interest: When do we move around the sun faster? In the daytime or at night?

A bit of a puzzler, isn't it? After all, it's always day on one side of the earth and night on the other. But don't dismiss my question as senseless. Note that I'm asking you not when the earth itself moves faster, but when we, who live on the earth, move faster in the heavens. And that is another pair of shoes.

In the solar system we make two motions; we revolve around the sun and simultaneously turn on the earth's axis. The two motions add, but with different results, depending whether we are on the daylit side or on the nightbound one.

Fig. 6 shows you that at midnight the speed of rotation is *added* to that of the earth's translation, while at noon it is, on the contrary, *subtracted* from the latter. Consequently, *at midnight we move faster in the solar system than at noon.* Since any point on the equator travels about half a kilometre a second, the difference there between midnight and midday speeds comes to as much as a whole kilometre a second.

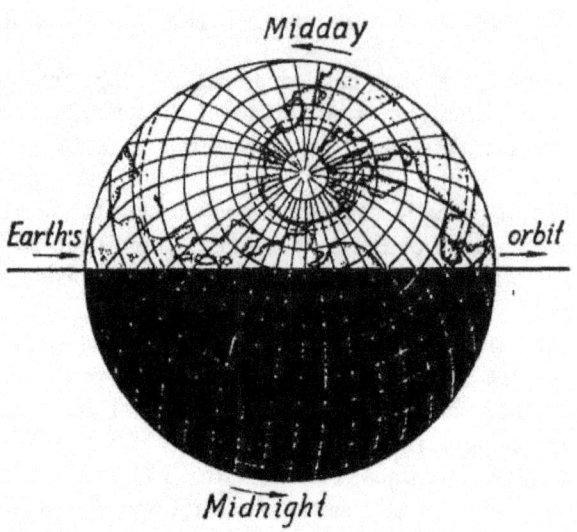

Midday

Earth's orbit

Midnight

Fig. 6. On the dark side we move around the sun faster
than on the sunlit side

Any of you who are good at geometry will easily reckon that for Leningrad, which is on the 60th parallel, this difference is only half as much. At 12 p.m. Leningraders travel in the solar system half a kilometre more a second than they would do at 12 a.m.

THE CART-WHEEL RIDDLE

Attach a strip of coloured paper to the side of the rim of a cart-wheel or bicycle tire, and watch to see what happens when the cart, or bicycle, moves. If you are observant enough, you will see that near the ground the strip of paper appears rather distinctly, while on top it flashes by so rapidly that you can hardly spot it.

Doesn't it seem that the top of the wheel is moving faster than the bottom? And when you look at the upper and lower spokes of the moving wheel of a carriage, wouldn't you think the same? Indeed, the upper spokes seem to merge into one solid body, whereas the lower spokes can be made out quite distinctly.

Incredibly enough, *the top of the rolling wheel does really move faster than the bottom.* And, though seemingly unbelievable, the explanation is a pretty simple one. Every point on the rolling wheel makes *two* motions simultaneously—one about the axle and the other forward together with the axle. It's the same as with the earth itself. The two motions add, but with different results for the top and bottom of the wheel. At the top the wheel's motion of rotation *is added* to its motion of translation, since both are in the same direction. At the bottom rotation is made in the *reverse* direction and, consequently, must be *subtracted* from translation. That is why the stationary observer sees the top of the wheel moving faster than the bottom.

A simple experiment which can be done at convenience proves this point. Drive a stick into the ground next to the wheel of a stationary vehicle opposite the axle. Then take a piece of coal or chalk and make two marks on the rim of the wheel—at the very top and at the very bottom. Your marks should be right opposite the stick. Now push the vehicle a bit to the right (*Fig. 7*), so that the axle moves some 20 to 30 cm away from the stick. Look to see how the marks have shifted. You will find that the upper mark *A* has shifted much further away than the lower one *B* which is almost where it was before.

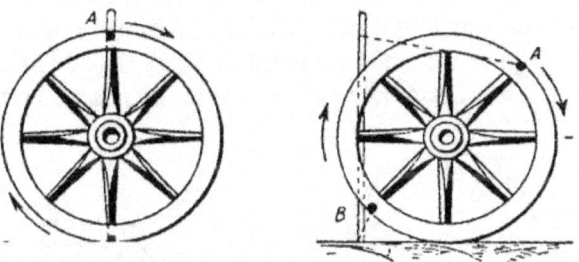

Fig. 7. A comparison between the distances away from the stick of points *A* and *B* on a rolling wheel (right) shows that the wheel's upper segment moves faster than its lower part

THE WHEEL'S SLOWEST PART

As we have seen, not all parts of a rolling cart-wheel move with the same speed. Which part is slowest? That which touches the ground. Strictly speaking, at the moment of contact, this part is absolutely stationary. This refers only to a rolling wheel. For the one that spins round a fixed axis, this is not so. In the case of a flywheel, for instance, all its parts move with the same speed.

BRAIN-TEASER

Here is another, just as ticklish, problem. Could a train going from Leningrad to Moscow have any points which, in relation to the railroad track, would be moving in the opposite direction? It could, we find. All the train wheels have such points every moment. They are at the bottom of the protruding rim of the wheel (the bead). When the train goes forward, these points move backward. The following experiment, which you can easily do yourself, will show you how this happens. Attach a match to a coin with some plasticine so that the match protrudes in the plane of the radius, as shown in *Fig. 8*. Set the coin together with the match in a vertical position on the edge of a flat ruler and hold it with your thumb at its point of contact—*C*. Then roll it to and fro. You will see that points *F*, *E* and *D* of the jutting part of the match

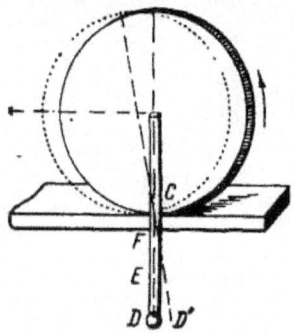

Fig. 8. When the coin is rolled leftwards, points *F*, *E* and *D* of the jutting part of the match move backwards

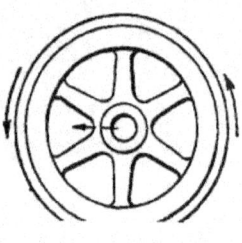

Fig. 9. When the train wheel rolls leftwards the lower part of its rim rolls the other way

Fig. 10. Top: the curve (a cycloid) described by every point on the rim of a rolling cart-wheel. Bottom: the curve described by every point on the rim of a train wheel

move not forwards but backwards. The further point D—the end of the match—is from the edge of the coin, the more noticeable backward motion is (point D shifts to D').

The points on the bead of the train wheel move similarly. So when I tell you now that there are points in a train that move not *forward* but *backward*, this should no longer surprise you. True, this backward motion lasts only the negligible fraction of a second. Still there is, despite all our habitual notions, a backward motion in a moving train. *Figs. 9* and *10* provide the explanation.

WHERE DID THE YACHT CAST OFF?

A rowboat is crossing a lake. Arrow a in *Fig. 11* is its velocity vector. A yacht is cutting across its course; arrow b is its velocity vector. Where did the yacht cast off? You would naturally point at once to point M. But you would get a different reply from the people in the dinghy. Why?

They don't see the yacht moving at right angles to their own course, because they don't realise that they are moving themselves. They think

Fig. 11. The yacht is cutting across the rowboat's course. Arrows *a* and *b* designate the velocities. What will the people in the dinghy see?

they're stationary, while everything around is moving with their own speed but in the opposite direction. From their point of view the yacht is moving not only in the direction of the arrow *b* but also in the direction of the dotted line *a*—opposite to their own direction (*Fig. 12*). The two motions of the yacht—the real one and the seeming one—are resolved according to the rule of the parallelogram. The result is that the people in the rowboat think the yacht to be moving along the diagonal of the parallelogram *ab*; that is also why they think the yacht cast off not at point *M*, but at point *N*, way in front of the rowboat (*Fig. 12*).

Travelling together with the earth in its orbital path, we also plot the position of the stars wrongly—just as the people in the dinghy did when asked where the yacht cast off from. We see the stars displaced slightly forward in the direction of the earth's orbital motion. Of course, the earth's speed is negligible compared with that of light (10,000

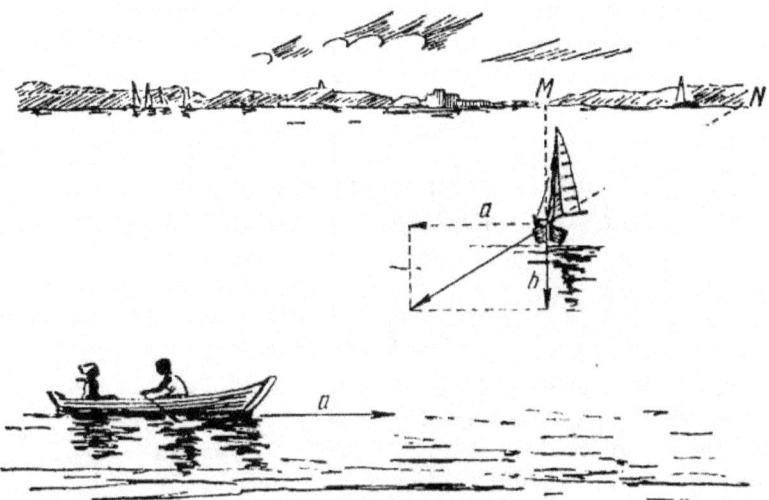

Fig. 12. The people in the dinghy think the yacht to be coming towards them slantwise—from point N

times less) and, consequently, this stellar displacement, known as aberration of light, is insignificant. However, we can detect it with the aid of astronomical instruments.

Did you like the yacht problem? Then answer another two questions related to the same problem. Firstly, give the direction in which the yachtsmen think the dinghy is moving. Secondly, say where the yachtsmen think the dinghy is heading. To answer, you must construct a parallelogram of velocities on the vector a (*Fig. 12*), whose diagonal will indicate that from the yachtsmen's point of view the dinghy seems to be moving slantwise, as if heading for the shore.

CHAPTER TWO

GRAVITY AND WEIGHT. LEVERS. PRESSURE

TRY TO STAND UP!

You'd think I was joking if I told you that you wouldn't be able to get up from a chair—provided you sat on it in a certain way, even though you wouldn't be strapped down to it. Very well, let's have a go. Sit down on a chair in the same way the boy in *Fig. 13* is sitting. Sit upright and *don't shove your feet under the chair*. Now try to get up without moving your feet or bending forward. You can't, however hard you try. You'll never stand up until you push your feet under the chair or lean forwards.

Before I explain, let me tell you about the equilibrium of bodies in general, and of the human body in particular. A thing will

Fig. 13. It's impossible to get up

not topple only when the perpendicular from its centre of gravity goes through its base. The leaning cylinder in *Fig. 14* is bound to fall. If, on the other hand, the perpendicular from its centre of gravity fell through its base, it wouldn't topple over. The famous leaning towers of Pisa and Bologna, or the leaning campanile in Arkhangelsk (*Fig. 15*), don't fall, despite their tilt, for the same reason. The perpendiculars from their centres of gravity do not lie outside their bases. Another reason is that their foundations are sunk deep in the ground.

You won't fall only when the perpendicular from your centre of gravity lies within the area bound by the outer edge of your feet (*Fig. 16*). That is why it is so hard to stand on one leg and still harder to balance on a tight-rope. Our "base" is very small and the perpendicular from the centre of gravity may easily come to lie outside its limits. Have you noticed the odd gait of an "old sea dog"? He spends most of his life aboard a pitching ship where the perpendicular from the centre of gravity of his body may come to fall outside his "base" any moment. That accustoms him to walk on deck so that his feet are set wide apart and take in as large a space as

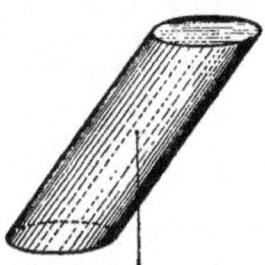

Fig. 14. The cylinder must topple as the perpendicular from its centre of gravity lies outside its base

Fig. 15. Arkhangelsk leaning campanile. A reproduction from an old photograph

possible, which saves him from falling. Naturally, he'll waddle in the same habitual fashion on hard ground as well.

Another instance—of an opposite nature this time. This is when the effort to keep one's balance results in a beautiful pose. Porters who carry loads on their heads are well-built—a point, I presume, you have noticed. You may have also seen exquisite statues of women holding jars on their heads. It is because they carry a load on their heads that these people have to hold their heads and bodies upright. If they were to lean in any direction, this would shift the perpendicular from the centre of gravity higher than usual, because of the head-load, outside the base and unbalance them.

Back now to the problem I set you at the beginning of the chapter. The sitting boy's centre of gravity is inside the body near the spine—

about 20 centimetres above the level of his navel. Drop a perpendicular from this point. It will pass through the chair behind the feet. You already know that for the man to stand up it should go through the area *taken up by the feet.* Consequently, when we get up we must either bend forward to shift the centre of gravity, or shove our feet beneath the chair to place our "base" below

Fig. 16. When one stands, the perpendicular from the centre of gravity passes through the area bound by the soles of one's feet

the centre of gravity. That is what we usually do when getting up from a chair. If we are not allowed to do this, we'll never be able to stand up—as you have already gathered from your own experience.

WALKING AND RUNNING

The things you do thousands of times a day, and day after day all your life, ought to be things you have a very good idea about, oughtn't they? Yes, you will say. But that is far from so. Take walking and running, for instance. Could anything be more familiar? But I wonder how many of you have a clear picture of what we really do when we walk and run, or of the difference between the two. Let's see what a physiologist has to say about walking and running. I'm sure most of you will find his description startlingly novel. (The passage is from Prof. Paul Bert, *Lectures on Zoology.* The illustrations are my own.)

"Suppose a person is standing on one leg, the right leg, for instance. Suppose further that he is lifting his heel, meanwhile bending forwards. [When walking or running a person exerts on the ground, when pushing his foot away from it, a pressure of some 20kg in addition to his weight. Hence a person exerts a greater pressure on the ground when he is moving than when standing.—*Y. P.*] In such a position the perpendicular from the centre of gravity will naturally be outside the base and the person is bound to fall forwards. Scarcely has he started doing this than he quickly throws forward his left leg, which was suspended thus far, to put it down on the ground in front of the perpendicular from the centre of gravity. The perpendicular thus comes to drop through the area bound by the lines linking the points of

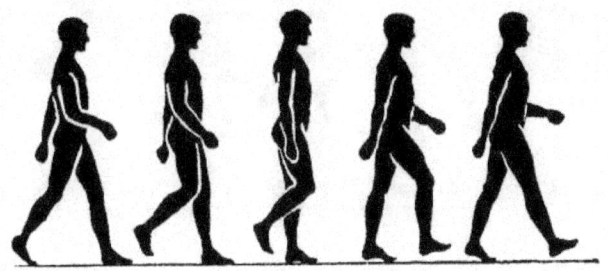

Fig. 17. How one walks. The series of positions in walking

support of both feet. Balance is thus restored; the person has taken a step forward.

"He may remain in this rather tiring position, but should he wish to continue forward, he will lean still further forward, shift the perpendicular from the centre of gravity outside the base, and again throw his leg—the right one this time—forwards when about to fall. He thus

B

Fig. 18. A graph showing how one's feet move when walking. Line *A* is the left foot and line *B* is the right foot. The straight sections show when the foot is on the ground, and the curves—when the foot is in the air. In the time-interval *a* both feet are on the ground; in the time-interval *b*, foot *A* is in the air and foot *B* still on the ground; in the timeinterval *c* both feet are again on the ground. The faster one walks, the shorter the time-intervals *a* and *c* get (compare with the "running" graph in *Fig. 20*).

takes another step forward. And so on and so forth. Consequently, walking is just *a series of forward fallings*, punctually forestalled by throwing the leg left behind into a supporting position.

Fig. 19. How one runs. The series of positions in running, showing moments when both feet are in the air

"Let's try to get to the root of the matter. Suppose the first step has already been made. At this particular moment the right foot is still on the ground and the left foot is already touching it. If the step is not very short the right heel should be lifted, because it is this rising heel that enables one to bend forward and change one's balance. It is the heel of the left foot that touches the ground first. When next the entire

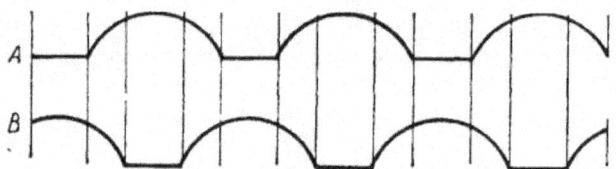

Fig. 20. A graph showing how one's feet move when running (compare with *Fig. 18*). There are time-intervals (*b, d* and *f*) when both feet are in the air. This is the difference between running and walking

sole stands on the ground, the right foot is lifted completely and no longer touches the ground. Meanwhile the left leg, which is slightly bent at the knee, is straightened by a contraction of the femoral triceps to become for an instant vertical. This enables the half-bent right leg to move forward without touching the ground. Following the body's movement the heel of the right foot comes to touch the ground in time for the next step forwards. The left leg, which at this moment has only the toes of the foot touching the ground and which is about to rise, goes through a similar series of motions.

"Running differs from walking in that the foot on the ground is energetically straightened by a sudden contraction of its muscles to throw the body forwards so that the latter is *completely off the ground for a very short interval of time*. Then the body again falls to come to rest on the other leg, which quickly moves forward while the body is still in the air. Thus, running consists of a series of *hops* from one foot to the other."

As for the energy a person expends in walking along a horizontal pavement it is not at all nil as some might think. With every step made, the centre of gravity of a walker's body is lifted by a few centimetres. A reckoning shows that the work spent in walking along a horizontal path is about a fifteenth of that required to raise the walker's body to a height equivalent to the distance covered.

HOW TO JUMP FROM A MOVING CAR

Most will surely say that one must jump forward, in the direction in which the car is going, in conformity with the law of inertia. But what does inertia have to do with it all? I'll wager that anyone you ask this question will soon find himself in a quandary, because according to inertia one should jump backwards, contrary to the direction of motion. Actually inertia is of secondary importance. If we lose sight of the main reason why one should jump forwards—one that has nothing to do with inertia—we will indeed come to think that we must jump backwards and not forwards.

Suppose you have to jump off a moving car. What happens? When you jump, your body has, at the moment you let go, the same velocity as the car itself—by inertia—and tends to move forwards. By jumping forwards, far from diminishing this velocity, we, on the contrary, increase it. Then shouldn't we jump *backwards*—since in that case the velocity thus imparted would be *subtracted* from the velocity our body possesses by inertia, and hence, on touching the ground, our body would have less of a toppling impetus?

But, when one jumps from a moving carriage, one always jumps forwards in the direction of its movement. That is indeed the best way, a time-honoured one, and I strongly warn you against trying to test the awkwardness of jumping backwards.

We seem to have a contradiction, don't we? Now whether we jump forwards or backwards we risk falling, since our bodies are still moving when our feet touch the ground and come to a halt. (See "When Is a Horizontal Line Not Horizontal?" from the third chapter of *Mechanics for Entertainment* for another explanation.) When jumping forwards, the speed with which our bodies move is even greater than when jumping backwards, as I have already noted. But it is much *safer* to jump forwards than backwards, because then we mechanically throw a leg forwards or even run a few steps, to steady ourselves. We do this *without thinking*; it's just like walking. After all, according to mechanics, walking, as was noted before, is nothing but a *series of forward fallings of our body, guarded against by the throwing out of a leg*. Since we don't have this guarding movement of the leg when falling *backwards* the danger is much greater. Then even if we do fall forwards we can soften the impact with our hands, which we can't do if we fall on our backs.

As you see, it is safer to jump forwards, not so much because of inertia, but because of ourselves. This rule is plainly inapplicable to *one's belongings*, for instance. A bottle thrown from a moving car forwards stands more chances of crashing when it hits the ground than if thrown backwards. So if you have to jump from a moving car and have some luggage with you, first chuck out the luggage *backwards* and then jump *forwards* yourself. Old hands like tramcar conductors and ticket inspectors often jump off stepping *backwards but with their backs turned to the direction in which they jump*. This gives them a double advantage: firstly they reduce the velocity that the body acquires by inertia, and, secondly, guard themselves against falling on their backs, as they jump with their faces forward, in the direction where they are most likely to fall.

CATCHING A BULLET

The following curious incident was reported during the First World War. One French pilot, while flying at an altitude of two kilometres, saw what he took to be a fly near his face. Trapping it with his hands, he was flabbergasted to find that he had caught a German bullet! How like the tall stories told by Baron Munchausen of legendary fame, who claimed he had caught cannon balls with bare hands! But there is nothing incredible in the bullet-catching story.

A bullet does not fly everlastingly with its initial velocity of 800-900 m/sec. Air resistance causes it to slow down gradually to a mere 40 m/sec towards the end of its journey. Since aircraft fly with a similar speed, we can easily have a situation when bullet and plane will be flying with the same speed, in which case the bullet, in its relation to the plane and its pilot, will be stationary or barely moving. The pilot can easily catch it with his hand, especially if gloved, because a bullet heats up considerably while whizzing through the air.

MELON AS BOMB

We have seen that in certain circumstances a bullet can lose its "sting". But there are instances when a gently thrown "peaceful" object has a destructive impact. During the Leningrad-Tiflis motor run in 1924, Caucasian peasants tossed melons, apples, and the like at the racing cars to express their admiration. However, these innocuous gifts made terrible dents and seriously injured the motorists. This happened because the car's velocity added to that of the tossed melons or apples, transforming them into dangerous projectiles. A ten-gramme bullet possesses the same energy of motion as a 4kg melon thrown at a car doing 120 km.p.h. Of course, the impact of a melon is not the same as the bullet's since melons, after all, are squashy.

When we have super-fast planes doing about 3,000 km.p.h.—a bullet's approximate velocity—their pilots may chance to encounter what we have just described. Everything in the way of a super-fast aircraft will ram into it. Machine-gun fire or just a chance handful of bullets dropped from another plane will have the same effect; these bullets will strike the aircraft with the same impact as if fired from a machine gun. Since the relative velocities in both cases are the same— the plane and bullet meet with a speed of about 800 m/sec—the destruction done when they collide is the same as well. On the contrary, bullets fired from behind at a plane moving with the same speed are harmless, as we have already seen.

Fig. 21. Water-melons tossed at a fast-moving car are as dangerous as bombs

In 1935 engine driver Borshchov prevented a railway disaster by cleverly taking advantage of the fact that objects moving in the same direction at practically the same speed come into contact without knocking each other to pieces. He was driving a train between Yelnikov and Olshanka, in Southern Russia. Another train was puffing along in front. The driver of this train couldn't work up enough steam to make the grade. He uncoupled his engine and several waggons and set off for the nearest station, leaving a string of 36 waggons behind. But as he did not place brake-shoes to block their wheels, these waggons started to roll back down the grade. They gathered up a speed of some 15 km. p.h. and a collision seemed imminent. Luckily enough, Borshchov had his wits about him and was able to figure out at once what to do. He braked his own train and also started a backward manoeuvre, gradually working up the same speed of 15 km.p.h. This enabled him to bring the 36 waggons to rest against his own engine, without causing any damage.

Finally this same principle is applied in a device making it easier for us to write in a moving train. You all know that this is hard to do because of the jolts when the train passes over the rail joints. They do not act simultaneously on both paper and pen. So our task is to

Fig. 22. Contraption for writing in a moving train

contrive something that would make the jolts act simultaneously on both. In this case they would be in a state of rest with respect to each other.

Fig. 22 shows one such device. The right wrist is strapped to the smaller board *a* which slides up and down in the slots in board *b*, which, in turn, slides to and fro along the grooves of the writing board placed on the train compartment table. This arrangement provides plenty of "elbow-room" for writing and at the same time causes each jolt to act simultaneously on both paper and pen, or rather the hand holding the pen. This makes the process as simple as writing on an ordinary table at home. The only unpleasant thing about it is that since the jolts again do not act simultaneously on both wrist and head, you get a jerky picture of what you're writing.

You will get your correct weight only if you stand on the scales without moving. As soon as you bend down, the scales show less. Why? When you bend, the muscles that do this also pull up the lower half of your body and thus diminish the pressure it exerts on the scales. On the contrary, when you straighten up, your muscles push the upper and lower halves of the body away from each other; in this case the scales will register a greater weight since the lower half of your body exerts a greater pressure on the scales.

You will change your weight-readings—provided the scales are sensitive enough—even by lifting an arm. This motion already slightly increases your body's seeming weight. The muscles you use to lift your arm up have the shoulder as their fulcrum and, consequently, push it together with the body down, increasing the pressure exerted on the scales. When you stop lifting your arm you start using another, opposite set of muscles; they pull the shoulder up, trying to bring it closer to the end of the arm; this reduces the weight of your body, or rather its pressure on the scales. On the contrary, when you lower your arm you reduce the weight of your body, to increase it when you stop lowering it. In brief, by using your muscles you can increase or reduce your weight, meaning of course the pressure your body exerts on the scales.

WHERE ARE THINGS HEAVIER?

The earth's pull diminishes the higher up we go. If we could lift a kilogramme weight 6,400 km up, to twice the earth's radius away from its centre, the force of gravity would grow $2^2 = 4$ times weaker, in which case a spring balance would register only 250 grammes instead of 1,000. According to the law of gravity the earth attracts bodies as if its entire mass were concentrated in the centre; the force of this attraction diminishes inversely to the square of the distance away. In our particular instance, we lifted the kilogramme weight twice the distance away from the centre of the earth; hence attraction grew $2^2 = 4$ times weaker. If we set the weight at a distance of 12,800 km away from the surface of the earth—three times the earth's radius—the force of attraction would grow $3^2=9$ times weaker, in which case our kilogramme weight would register only 111 grammes on a spring balance.

You might conclude that the deeper down in the earth we were to put our one-kilogramme weight, the greater the force of attraction would grow and the more it should weigh. However, you would be mistaken. The weight of a body does not increase; on the contrary, it diminishes.

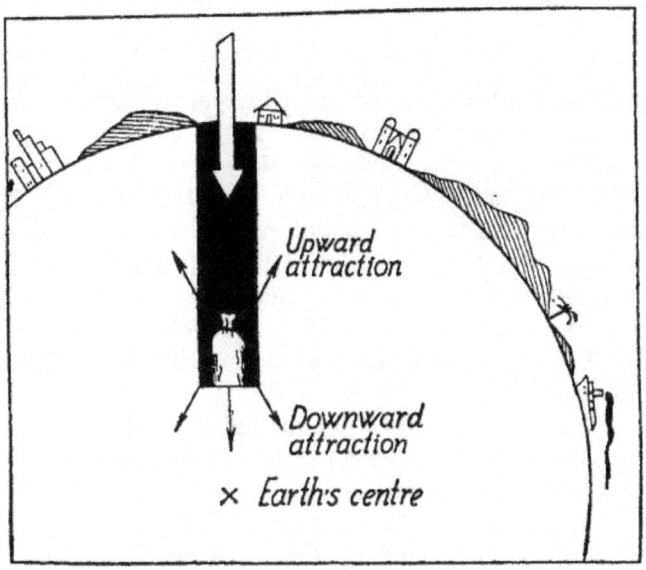

Fig. 23. Gravitational pull lessens the closer we get to the
middle of the Earth

This is because now the earth's attracting forces no longer act just on one side of the body but all around it. Fig. 23 shows you the weight in a well; it is pulled down by the forces below it and simultaneously up by the forces above it. It is really only the pull of that spherical part of the earth, the radius of which is equal to the distance from the centre of the earth to the body, that is of importance. Consequently, the deeper down we go, the less a body should weigh. At the centre of the earth it should weigh nothing, as here it is attracted by equal forces on all sides.

To sum up: a body weighs most at the earth's surface; its weight diminishes whether it is lifted up from the earth's surface or interred (this would stand, naturally, only if the earth were homogeneous in density throughout). Actually, the closer to its centre, the greater the earth's density; at first the force of gravity grows to some distance down; only then does it start to diminish.

HOW MUCH DOES A FALLING BODY WEIGH?

Have you noticed that odd sensation you experience when you *start* to go down in a lift? You feel abnormally light; if you were falling into a bottomless abyss you would feel the same. This sensation is caused by weightlessness. At the very first moment when the lift-cabin floor has already started to go down but you yourself have still not acquired its velocity, your body exerts scarcely any pressure at all on the floor, and, consequently, *weighs* very little. An instant later this queer sensation is gone. Now your body seeks to fall faster than the smoothly running lift; it exerts a pressure on the cabin floor, reacquiring its full weight.

Tie a weight to the hook of a spring balance and observe the pointer as you quickly lower the balance together with the weight. For convenience's sake insert a small piece of cork in the slot and observe how it moves. The pointer will fail to register the full weight; it will be much less! If the balance were falling freely and you would be able to watch its pointer meanwhile, you would see it register a zero weight.

The heaviest object will lose all its weight when falling. The reason is simple. "Weight" is the force with which a body pulls at something holding it up or presses down on something supporting it. A *falling* body cannot pull the balance spring as it is falling together with it. A falling body does not pull at anything or press down on anything. Hence, to ask how much something weighs when falling is the same as to ask how much it weighs when it does not weigh.

Galileo, the father of mechanics, wrote way back in the 17th century in his *Mathematical Proofs Concerning Two Fields of a New Science*: "We feel a load on our back when we try to prevent it from dropping. But if we were to drop as fast as the load does, how could it press upon and burden us? This would be the same as to try to transfix with a spear [without letting go of it—*Y. P.*] somebody running ahead of us as fast as we are running ourselves."

The following simple experiment well illustrates this point. Place a nutcracker on one of the scale pans, with one arm on the pan and the

Fig. 24. Falling bodies are weightless

other tied by a piece of thread to the hook of the scale arm (*Fig. 24*). Add weights to the other pan to balance the nutcracker. Apply a lighted match to the thread. The thread will burn through and the suspended nutcracker arm will fall onto the pan. Will the pan holding the nutcracker dip? Will it rise? Or will it remain in equilibrium? Since you know by now that a falling body weighs nothing, you should be able to give the correct answer. The pan will rise for a moment. Indeed, though joined to the lower arm the nutcracker's upper arm nevertheless exerts less of a pressure on the pan when falling than when stationary. For a moment the nutcracker's weight diminishes, and thus the pan holding it rises.

FROM EARTH TO MOON

The years between 1865 and 1870 saw the publication in France of Jules Verne's *From the Earth to the Moon*, in which he set forth a fantastic scheme to shoot at the Moon an enormous projectile with people inside. His description seemed so credible that most of you who have read this book have probably hazarded whether this really could be done. Well, let's discuss it. (Today, after Sputnik and Lunik, we know that it is rockets, not cannon projectiles, that will be used for space travel. However, since a rocket flies after its last engine burns out, in accord with the same laws of ballistics, don't think Perelman is behind the times.)

Let's see at first whether we can fire a shell from a gun—at least theoretically—so that it never falls back to earth again. Theory tells us that it's possible. Indeed, why does a shell fired horizontally eventually fall back on earth again? Because the earth attracts it, curving its trajectory. Instead of keeping up a straight course, it curves towards the ground and is, therefore, bound to hit it sooner or later. The earth's surface is also curved, but the shell's trajectory is bent still more. However, if we made the shell follow a trajectory curved in exactly the same way as the earth's surface it would never fall back on earth again. Instead, it would trace an orbit concentric with the earth's circumference, becoming its satellite, a baby moon.

But how are we to make the shell follow such a trajectory? All we must do is to impart a sufficient initial velocity. Look at *Fig. 25* which depicts a cross-section of part of the earth. A cannon is mounted on the hilltop at point *A*. A shell fired horizontally from it would reach point *B* a second later—if not for the earth's gravitational pull. Instead, it reaches point *C* five metres lower than *B*. Five metres is the distance any freely falling body travels (in a void) in the first second — due to earth's surface gravitational pull. If, after it drops these five metres, our shell is at exactly the same distance away from the ground as it was when fired at point *A*, it means that the shell is

Fig. 25. How to reckon a projectile's "escape" velocity

following a trajectory curved concentrically to the earth's circumference.

All that remains is to reckon the distance AB (*Fig. 25*), or, in other words, the distance the shell travels horizontally in the space of a second, which will tell us the speed we need. In the triangle AOB, the side OA is the earth's radius (roughly 6,370,000 m); $OC=OA$ and $BC=5$m; h nce OB is 6,370,005 m. Applying Pythagoras's theorem we get:

$$(AB)^2=(6,370,005)^2-(6,370,000)^2.$$

We resolve this equation to find AB equal to roughly 8 km.

So, if there were no drag a shell shot horizontally with a muzzle velocity of 8 km/sec *would never fall back to earth again*; it would be an everlasting baby moon.

Now suppose we imparted to our shell a still greater initial velocity. Where would it fly then? Scientists dealing with celestial mechanics have proved that velocities of 8, 9 and even 10 km/sec give a trajectory shaped like an ellipse which would be the more elongated the greater the initial speed is. When the velocity reaches 11.2 km/sec, the shell will describe not an ellipse but a non-locked curve, a parabola, and fly away from the earth never to return (*Fig. 26*). So, *theoretically* it is quite possible to fly to the Moon inside a cannon ball, provided its muzzle speed is big enough. This, however, is a problem that may

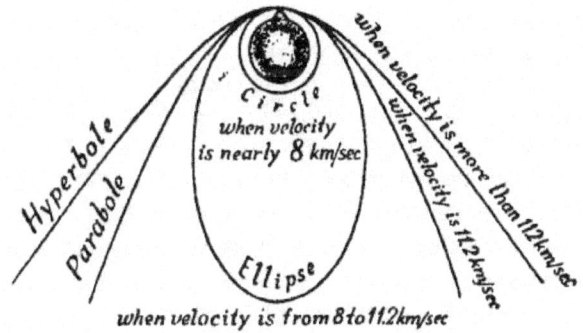

Fig. 26. When a projectile is fired with a starting velocity
of 8 km/sec and more

present some quite specific difficulties. Let me refer you, for greater detail, to Book Two of *Physics for Entertainment* and also to *Interplanetary Travel*—another book of mine. (In the foregoing we dismissed the drag which in real life would exceedingly complicate the attainment of such great velocities and perhaps render the task absolutely impossible.)

FLYING TO THE MOON: JULES VERNE VS. THE TRUTH

Any of you who have read *From the Earth to the Moon* most likely remembers the interesting passage describing the projectile's intersection of the boundary where the Moon matches the Earth in attraction. Wondrous things happened. All the objects inside the projectile became weightless; the travellers themselves began to float in the air.

There is nothing wrong in all this. What Jules Verne did lose sight of was that this happens not only at the point the novelist gave. It happens before and after as well—in fact, *as soon as free flight begins.*

It seems incredible, doesn't it? I'm sure though that soon you will be surprised not to have noticed this signal omission before. Let's turn to Jules Verne for an example. You haven't forgotten how the space travellers ejected the dead dog and how surprised they were to see it continue to trail behind the projectile instead of falling back to earth. Jules Verne described and explained this correctly. In a void all bodies fall with the same speed, with gravity imparting an identical acceleration to each. So, owing to gravity, both the projectile and the dead dog should have acquired the same falling velocity (an identical acceleration). Rather should we say that due to gravity their starting velocities diminished in the same measure. Consequently, both should whizz along with the same velocity; that is why after its ejection the dead dog kept on trailing along in the projectile's wake.

Jules Verne's omission was: if the dead dog did not fall back to earth again *after the ejection*, why should it fall when *inside* the projectile? The same forces act in both cases! The dead dog suspended in mid-air inside the projectile should remain in that state as its speed is absolutely the same as the projectile's; hence it is in a state of rest *in respect to the projectile.*

44

What goes for the dead dog also goes for the travellers and all objects, in general, inside the projectile, as they all fly along the trajectory with the same speed as the projectile and should not fall, even though having nothing to stand, sit, or lie on. One could take a chair, turn it upside down and lift it to the ceiling; it won't fall "down", because it will go on travelling together with the ceiling. One could sit on this chair also upside down and not fall either. What, after all, could make him fall? If he did fall or float down, this would mean that the projectile's speed would be greater than that of the man on the chair; otherwise the chair wouldn't float or fall. But this is impossible since we know that everything inside the projectile has the same acceleration as the projectile itself. This was what Jules Verne failed to take into account. He thought everything inside the projectile would continue to press down on its floor when it was in space. He forgot that a weight presses down on what supports it only because this support is stationary. But if both object and its support hurtle with the same velocity in space they simply can't press down on each other.

So, as soon as the projectile began to fly further on by its own momentum, its travellers became completely weightless and could float inside it, just as everything else could, too. That alone would have immediately told the travellers whether they were hurtling through space or still inside the cannon. Jules Verne, however, says that in the first half hour after the projectile was shot into space they couldn't guess whether they were moving or not, however hard they tried.

"'Nicholl, are we moving?'

"Nicholl and Barbicane looked at each other; they had not yet troubled themselves about the projectile.

"'Well, are we really moving?' repeated Michel Ardan.

"'Or quietly resting on the soil of Florida?' asked Nicholl.

"'Or at the bottom of the Gulf of Mexico?' added Michel Ardan."

These are doubts a steamboat passenger may entertain; they are absolutely out of the question for a space traveller, because he can't help noticing his complete loss of weight, which the steamboat passenger naturally retains.

Jules Verne's projectile must certainly be a very queer place, a tiny world of its own, where things are weightless and float and stay where they are, where objects retain their equilibrium wherever they are placed, where even water won't pour out of an inclined bottle. A pity Jules Verne slipped up, when this offers such a delightful opportunity for fantasy to run riot! (If this problem interests you, we could refer you to the appropriate chapter in A. Sternfeld's *Artificial Earth Satellites*.)

FAULTY SCALES CAN GIVE RIGHT WEIGHT

What is more important to get the right weight—scales or weights? Don't think both identically important. You can get the right weight even on faulty scales as long as you have the right weights. Of the several methods used, we shall deal with two.

One was suggested by the great Russian chemist Dmitry Mendeleyev. You begin by placing anything handy on one of the pans. Make sure that it is heavier than the object you want to weigh. Balance it with weights on the other pan. Then place what you want to weigh on the pan holding the weights and remove the necessary number of weights to bring to balance again. Tote up the weights removed to get the weight of what you wanted to weigh. This is called "the constant load method" and is particularly convenient when several objects need to be weighed in succession. The initial load is used to weigh everything you have to weigh.

Another method, called the "Borda method" after the scientist who proposed it, is as follows:

Place the object you want to weigh on one of the pans. Then pour sand or shot into the other pan till the scales balance. Remove your object from the pan—but don't touch the sand or shot in the other pan!—and place weights in the emptied pan till the scales balance again. Tote up these weights to find how much your object weighs. This is also called "replacement weighing".

This simple method can also be used for a one-pan spring balance, provided of course you have correct weights. In this case you don't need either sand or shot. Just put your object on the pan and note the reading. Then remove the object and place in the pan as many weights as needed to get the same reading. Their combined weight will give the weight of the object they replace.

How much can you lift with one arm? Let's say it's ten kilogrammes. Does this amount qualify your arm's muscle-power? Oh, no. Your biceps is much stronger. *Fig. 27* shows how this muscle works. It is attached close to the fulcrum of the lever that the bone of your forearm represents. The load you are lifting acts on the other end of this live lever. The distance between the load and the fulcrum, that is, the joint, is almost eight times more than that between the end of the biceps and the fulcrum. This means that if you are lifting a load of 10 kg your biceps is exerting eight times as much power, and, consequently, could lift 80 kg.

It would be no exaggeration to say that everybody is much stronger than he is, or rather that one's muscles are much more powerful than what we can really do with them. Is this an expedient arrangement? Not at all, you might think at first glance. We seem to have totally unrewarded loss. Recall, however, an old "golden rule" of mechanics: whatever you lose in power you gain in displacement. Here you gain in speed; your arm moves eight times faster than its muscles do. The muscular arrangement in animals enables them to move extremities quickly, which is more important than strength in the struggle to survive. Otherwise, we would move around at literally a snail's pace.

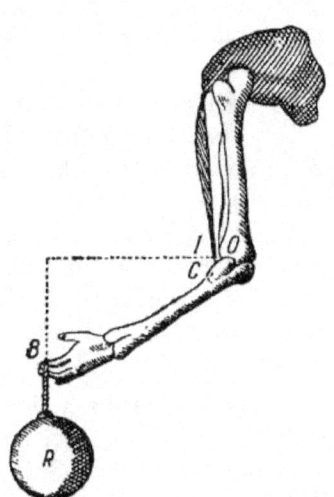

Fig. 27. Forearm *C* acts as a lever. The force acts on point *I*; the fulcrum is at point *O* and the load *R* is being lifted from point *B. BO* is roughly eight times longer than *IO*. (This drawing is from an ancient book called *Concerning the Motions of Animals* by the 17th-century Florentine scholar Borelli who was the first to apply the laws of mechanics to physiology.)

WHY DO SHARP THINGS PRICK?

Have you ever wondered why a needle so easily pierces things? Why is it so easy to drive a needle through a piece of cloth or cardboard and so hard to do the same thing with a blunt nail? After all, doesn't the same force act in both cases? The force is the same, but the *pressure* isn't. In the case of the needle the entire force is concentrated on its point; in the case of the nail the same amount of force is distributed over the larger area of the blunt end. So, though we exert the same force, the needle gives a much greater pressure than the blunt nail.

You all know that a twenty-toothed harrow loosens the soil more deeply than a sixty-toothed one of the same weight. Why? Because the *load on each tooth* of the first harrow is more than on each tooth of the second.

When we speak of pressure, we must always take into consideration, besides force, also the area upon which this force acts. When we are told that a worker is paid a hundred rubles, we don't know whether this is much or little, because we don't know whether this is for a whole year or for just one month.

Similarly does the action of a force depend on whether it is distributed over a square centimetre or concentrated on the hundredth of a square millimetre. Skis easily take us across fresh snow; without them we fall through. Why? On skis the weight of your body is distributed over a much greater area. Supposing the surface of our skis is 20 times more than the surface of our soles, on skis we would exert on the snow a· pressure which is only a twentieth of the pressure we exert when we have no skis on. As we have noticed, fresh snow will bear you when you are on skis, but will treacherously let you down when you're without them.

For the same reason horses used in marshlands are shod in a special fashion giving them a wider supporting area and lessening the pressure exerted per square centimetre. For the same reason people take the same precautions when they want to cross a bog or thin ice, often crawling to distribute their weight over a greater area.

Finally, tanks and caterpillar tractors don't get stuck in loose ground, though they are very heavy, again because their weight is distributed over a rather great supporting area. An eight-ton tractor exerts a pressure of only 600 grammes per square centimetre. There are caterpillars which exert a pressure of only 160 gr/cm² despite a two-ton load, which makes for the easy crossing of peatbogs and sand-beaches. Here it is a large supporting area which gives the advantage, whereas in the case of the needle it is the other way round.

This all shows that a sharpened edge pierces things only because it has a very minute area for the force to act upon. That is why a sharp knife cuts better than a blunt one: the force is concentrated on a smaller area of the knife edge. To sum up: sharp objects prick and cut well, because much pressure is concentrated on their points and edges.

COMFORTABLE BED ... OF ROCK

Why is it pleasanter to sit on a chair than on a flat-topped stool though both are of wood? Why is it pleasant to lie in a hammock though the pieces of rope that go to make it are by no means soft?

I suppose you've already guessed why. The stool-top is flat; when you sit on it, you press down with your entire weight on a small area. Chairs, on the other hand, usually have a concave seat; in this case you press down on a much greater area, over which your weight is distributed. To every unit of surface you have a smaller weight, smaller pressure.

The trick, as you see, is to distribute pressure more evenly. On a soft bed we make depressions that conform to the uneven shape of our bodies. Pressure is distributed rather evenly, with only a few grammes per square centimetre. No wonder we find it so pleasant.

The following reckoning well illustrates the difference. An adult person has a body surface of about 2m², or 20,000 cm². In bed roughly a quarter of it—0.5 m², or 5,000 cm²—supports him. Presuming that he weighs about 60 kg, or 60,000 gr, this would mean that we have a pressure of only 12 gr/cm². On bare boards he would have a supporting area of only some 100 cm². There are fewer points of contact. This means a pressure per sq. cm. of half a kilogramme instead of a dozen grammes. Quite a noticeable difference, isn't it? And one feels it at once.

But even the *hardest* of beds would be as soft as eiderdown, provided the weight of your body were distributed all over it. Suppose you left the imprint of your body in wet clay. When it hardens—drying clay shrinks by some five to ten per cent, but we shall discount this—you could lie in it again and think yourself in a featherbed. Though you would be lying on what is practically rock, it would feel soft, because your weight would be distributed over a much greater area of support.

CHAPTER THREE

ATMOSPHERIC RESISTANCE

BULLET AND AIR

Every schoolboy knows that the air impedes a bullet in its flight. Few, however, know what a great impediment it is. Most think such a "caressing" environment as the air—which is something we usually never feel—could not really get in the way of a fast-flying rifle bullet.

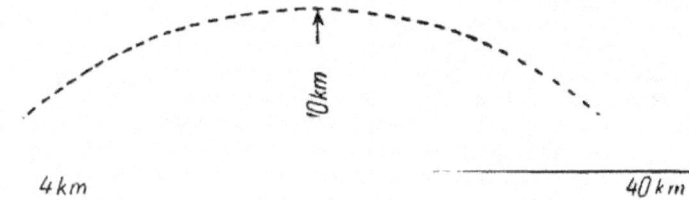

Fig. 28. Flight of a bullet in the air and in a vacuum. The big arc is the trajectory described when there is no atmosphere. The tiny, left-hand arc is the real trajectory

However, one good glance at *Fig. 28* will already make you realise that the air places quite a serious obstacle in the bullet's way. The large curve on the diagram designates the trajectory the bullet would describe were there no air. In this case, after flying out of a rifle tilted at 45°, and with an initial velocity of 620 m/sec, the bullet would describe a vast arc ten kilometres high and fly almost 40 km. But actually our bullet flies only 4 km, describing the tiny arc which is scarcely noticeable side by side with the first one. That is what the resistance of the air, the air drag, does!

BIG BERTHA

The Germans were the first—in 1918, towards the close of the First World War, when French and British aircraft had put a stop to German air raids—to practise long-range artillery bombardment from a distance of 100 kilometres and more.

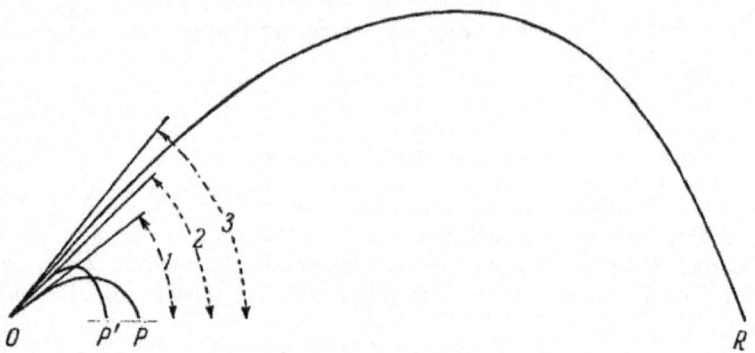

Fig. 29. The range changes when the mouth of a long-distance gun is tilted at different angles. In the case of angle *1*, the projectile strikes *P*, and in the case of angle *2*, *P'*, but in the case of angle *3*, it flies much farther as it goes through the rarefied stratosphere

It was by chance that German gunners hit upon their absolutely novel method for shelling the French capital, which was then at least 110 km away from the front lines. Firing shells from a big cannon tilted up at a wide angle, they unexpectedly discovered that they could make them fly 40 km instead of 20. When a shell is fired steeply upwards with a great initial velocity, it reaches a high-altitude, rarefied atmospheric strata, where the air drag is rather weak. Here it flies for quite a distance, before veering steeply to fall back to earth again. *Fig. 29* illustrates the great difference in trajectory at different angles of the gun barrel. This became the basic principle of the long-range gun that the Germans designed to bombard Paris from 115 km away. Such a gun was made—Big Bertha—and it fired more than 300 shells at Paris throughout the summer of 1918.

It was learned later that Big Bertha consisted of a tremendous steel tube 34 metres long and 1 metre thick. The breech walls were 40 cm thick. The gun itself weighed 750 tons. Its 120 kg shells were one metre long and 21 cm thick. Each charge took 150 kg of gunpowder which developed a pressure of 5,000 atmospheres, ejecting the shell with an initial velocity of 2,000m/sec. Since the angle of elevation was 52°, the shell described a tremendous arc, reaching its highest point way up in the stratosphere 40 km above the ground. It took the shell only 3.5 minutes to reach Paris, 115 km away; two minutes were spent in the stratosphere.

Fig. 30. Big Bertha

Big Bertha was the first long-range gun in history, the progenitor of modern long-range artillery.

Let me note that the greater the initial velocity of a bullet or shell, the more resistance the air puts up, increasing, moreover, in proportion to the square, cube, etc., of the velocity, depending on its amount.

WHY DOES A KITE FLY?

Do you know why a kite soars when pulled forward by the twine? If you do, you will also be able to understand why airplanes fly and maple seeds float. You'll even be able to fathom to some extent the causes of the boomerang's very odd behaviour. Because all these things are related. The very same air which is so great an impediment to a bullet or a shell enables the light maple seed to float and even heavy airliners to fly.

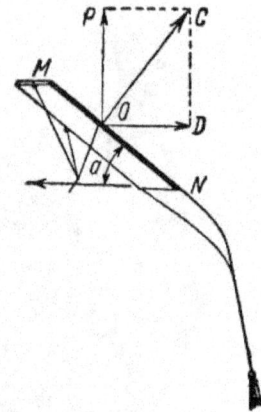

Fig. 31. The forces that make a kite fly

If you don't know why a kite flies, the simple drawing in *Fig. 31* will provide the explanation. Let line *MN* designate the kite's cross-section. When you let the kite go and pull at the cord, the kite, because of its heavy tail, moves at an angle to the ground. Let the kite move from right to left and *a* be the angle at which the plane of the kite is inclined to the horizon. We shall now proceed to examine the forces that act on the kite. The air, of course, should obstruct its movement and exert some pressure on it, designated on *Fig. 31* by the vector *OC*. Since the air always presses perpendicular to the plane, *OC* is at right angles to *MN*. The force *OC* may be resolved into two forces by constructing what is called a *parallelogram of forces*. This gives us the two forces *OD* and *OP*. Of these two, the force *OD* pushes the kite back, thus reducing its initial velocity. The other force, *OP*, pulls the kite up, reducing its weight. When this force is big enough it overcomes the weight of the kite and lifts it. That is why the kite goes *up* when you pull it *forwards*.

The airplane is also a kite really, with the difference that its forward motion, which makes it go up, is imparted not by our pulling at it but by the propeller or jet engine. This is, of course, a very crude explanation. There are other factors that cause an airplane to rise. They are explained in Book Two of *Physics for Entertainment* under the heading "Waves and Whirlwinds".

LIVE GLIDERS

As you see aircraft are not made like birds, as one usually thinks, but rather like flying squirrels or.flying fish, which, by the way, employ their flying mechanism not to fly up but merely to take rather big leaps—or what a flier would call "glides". In their case, the force *OP*

(*Fig. 31*) is too small to offset their weight; it merely reduces their weight, enabling them to make very big jumps from some high point (*Fig. 32*). A flying squirrel can jump 20-30 m from the top of one tree to the lower branches of another. In the East Indies and in Ceylon a much larger species of flying squirrel is found. This is the kaguan, a flying lemur, which is about the size of our house cat and which has a wing spread of about half a metre, enabling it to leap some 50 m, despite its great weight. As for the phalangers that inhabit the Sunda Isles and the Philippines, they can jump as far as 70 m.

Fig. 32. Flying squirrels jump from 20 to 30 m

BALLOONING SEEDS

Plants also often employ a gliding mechanism to propagate. Many seeds have either a parachuting tuft or hairy appendages (the pappus), as in dandelions, cotton balls, and "goat's beards", or "wings", as in conifers, maples, white birches, elms, lindens, many kinds of umbelliferae, etc.

In Kerner von Marilaum's well-known *Plant Life*, we find the following relevant passage:

"On windless sunny days a host of seeds and fruits are lifted high up by vertical air currents. However, after dusk they usually float down a short cry away. It is important for seeds to fly, not so much to cover a wide area as to inhabit cracks in terraces and cliffs, which they would never reach in any other way. Meanwhile, horizontal air currents may carry these hovering seeds and fruits rather far.

"The seeds of some plants retain their wings and parachutes only while they fly. Thistle seeds quietly float until they encounter an

55

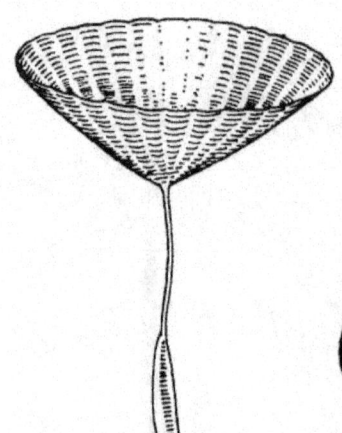

obstacle, when the seed discards its parachute and drops to the ground. That is why we see the thistle so often near walls and fences. But there are other cases, when the seed is attached permanently to its parachute."

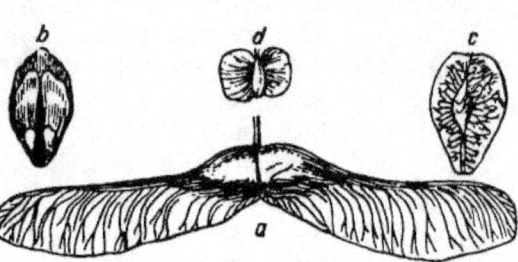

Fig. 33. Fruit of "goat's beard"

Fig. 34. Winged seeds of a) maple, b) pine-tree, c) elm, and d) birch

Figs. 33 and 34 show some seeds and fruits that have a gliding mechanism. As a matter of fact these plant "gliders" beat man-made ones on many points. They can lift a load which may be much greater than their own weight and automatically stabilise it. Thus if the seed of the Indian jasmine should chance to turn over, it will automatically regain its initial position with its convex side bottom-most, but when it meets an obstacle it doesn't capsize and drop like a plummet, but coasts down instead.

DELAYED PARACHUTE JUMPING

This, naturally, brings to mind the brave jumps parachutists sometimes make. They bail out at altitudes of some ten kilometres and pull the ripcord only after plummeting like a stone without opening their parachutes for quite a distance. Many think that in this delayed jump the parachutist falls as if in empty space. If this were really so, the delayed jump would be a much shorter affair, while the near-ground velocity would be tremendous.

However, atmospheric resistance prevents acceleration. The velocity of the falling parachutist during a delayed jump increases only in the first ten seconds, only for the first few hundred metres. Meanwhile atmospheric resistance increases, to finally reach a point where all further acceleration stops and the falling becomes even.

Here is a crude idea of a delayed jump from the angle of mechanics. *Acceleration* continues for only the first 12 seconds or even less, depending on the parachutist's weight. In this period he drops some 400-450 m and works up a velocity of about 50 m/sec. After that he falls uniformly, with the same speed, until he pulls the ripcord. Raindrops fall similarly. The only difference is that the initial period of acceleration for the raindrop is no more than a second. Consequently its near-ground velocity is not so great as in a delayed parachute jump, being between 2 and 7 metres a second, depending on its size. (Read my *Mechanics for Entertainment* for more about raindrop velocity and my *Do You Know Your Physics?* for more about delayed parachute jumping.)

THE BOOMERANG

For long this ingenious weapon, the most perfect technical device primitive man ever invented, had scientists wonderstruck. Indeed, the queer tangled trajectory the boomerang traces (*Fig. 35*) can tease any mind. Nowadays we have an elaborate theory to explain the boomerang; it is no longer a wonder. This theory is too intricate to explain at length. Let me merely note that boomeranging is the combined result of three factors: firstly, the initial throw; secondly, the boomerang's own rotation, and thirdly, atmospheric resistance. The Australian aborigine instinctively knows how to combine all three, deftly changing the boomerang's tilt and direction, and he throws it with a greater or smaller force to obtain the desired result.

You, too, can acquire some knack in boomerang-throwing. To make one for indoors, cut it out of cardboard, in the form shown in *Fig. 36*. Each arm is about 5 cm long and a little less than a centimetre

Fig. 35. Australian aborigine throwing a boomerang. The dotted line shows the trajectory of the boomerang, should it miss its target

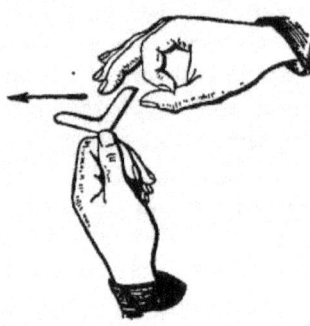

Fig.36. A cardboard boomerang and how to "throw" it

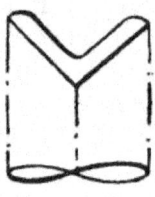

Fig. 37. Another cardboard boomerang (real size)

wide. Press it under the nail of your thumb and flick it forwards and a bit upwards. It will fly some five metres, loop, and return to your feet, provided it doesn't hit anything on the way. You can make a still better boomerang by copying the one given in *Fig. 37*, and also by twisting it to look somewhat like a propeller (as shown at the bottom of *Fig. 37*). After some experience you should be able to make it describe intricate curves and loops before it returns to your feet.

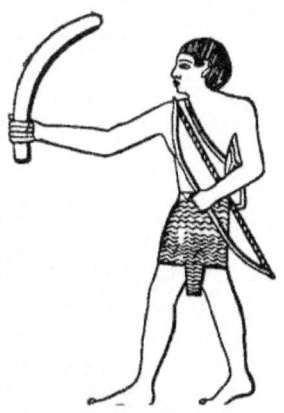

Fig. 38. Ancient Egyptian warrior throwing a boomerang

In conclusion let me note that the boomerang is not at all exclusively an Australian missile as is usually thought. It was employed in India and according to extant murals it was once commonly used by Assyrian warriors (see *Fig. 38*). It was also familiar in ancient Egypt and Nubia. The Australian boomerang's only distinguishing feature is the propeller-like twist that we mentioned, sending it into such a maze of whirls and loops, returning it to the thrower, *should he miss.*

CHAPTER FOUR

ROTATION. "PERPETUAL MOTION" MACHINES

HOW TO TELL A BOILED AND RAW EGG APART?

How can we find out whether an egg is boiled or not, without breaking the shell?

Mechanics gives us the answer. The whole trick is that a boiled egg spins differently than a raw one. Take the egg, place it on a flat plate and twirl it (*Fig. 39*). A cooked egg, especially a hard-boiled one, will revolve *much faster and longer* than a raw one; as a matter of fact, it is hard even to make the raw egg turn. A hard-boiled egg spins so quickly that it takes on the hazy form of a flat white ellipsoid. If flicked sharply enough, it may even rise up to stand on its narrow end.

The explanation lies in the fact that while a hard-boiled egg revolves as one whole, a raw egg doesn't; the latter's liquid contents do not

Fig. 39. Spinning an egg

Fig. 40. Telling a boiled egg from a raw one.

have the motion of rotation imparted at once and so act as a brake, retarding by force of inertia the spinning of the solid shell. Then boiled and raw eggs stop spinning differently. When you touch a twirling boiled egg with a finger, it stops *at once*. But a raw egg will resume spinning for a while after you take your finger away. Again the force of inertia is responsible. The liquid contents of the raw egg still continue moving after the solid shell is brought to a state of rest. Meanwhile the contents of the boiled egg stop spinning together with the outer shell.

Here is another test, similar in character. Snap rubber bands around a raw egg and a boiled one, along their "meridian", as it were, and hang them up by two identical pieces of string (*Fig. 40*). Twist the strings. giving the same number of turns, and then let them go. You will spot the difference between the two eggs at once. Inertia causes the boiled egg to overshoot its starting position and give the string some more twists in the opposite direction; then the string unwinds again with the egg again giving several turns; this continues for some time, the number of twists gradually diminishing until the egg comes to rest. The raw egg, on the other hand, scarcely overshoots its initial position at all; it will give but one or two turns and stop long before the boiled egg does. As we already know, this is due to its liquid contents which impede its movement.

WHIRLIGIG

Open an umbrella, stand it up with its top on the floor and twist the handle. You can easily make it revolve rather quickly. Now throw a little ball or a crumpled piece of paper into the umbrella. It won't stay there; it will be shot out by what has wrongly come to be called the "centrifugal force" but which is actually nothing but a manifestation of the force of inertia. The ball or piece of paper will be thrown off, not along the continuation of the radius but at a tangent to the circular motion.

At some public parks one may find an amusement (*Fig. 41*) based on this principle of rotation, where you may try out the law of inertia on yourself. This is a sort of whirligig with a round floor on which people either stand, sit, or lie. A concealed motor starts the floor revolving,

Fig. 41. A whirligig. Centrifugal forces are hurling the boys off

increasing its speed till inertia makes everybody on it slither or slide towards its edge. At first this is hardly noticeable, but the further away one gets from the centre, the more noticeable do both speed and, consequently, inertia grow. You try hard to hold on, but it is to no avail and finally you are hurled off.

The Earth itself is, in point of fact, a huge whirligig. Though it doesn't hurl us off, it does reduce our weight. At the equator, where rotation is fastest, one can "shed" a 300th of one's weight in this manner. This, plus another factor, the Earth's compression, reduces weight at the equator by about 0.5% or 1/200th. An adult person will consequently weigh 300 grammes less at the equator than at any of the poles.

INKY WHIRLWINDS

Make a teetotum, as shown in life size in *Fig. 42*, out of white cardboard and a match sharpened at one end. No particular knack is needed to twirl it—it's something any child can do. But though a child's toy, it can be very instructive. Do the following. Spill a few drops of ink on it and set it spinning before the ink dries. When it stops, look

Fig. 42. Ink drop traces on a twirling teetotum

to see what has happened to the ink drops. They will have drawn whorls—a miniature whirlwind.

Incidentally, this resemblance is not accidental. The whorls on the teetotum trace the movement of the ink drops, which undergo exactly what you experienced on the revolving floor. As the drop shoots away from the centre due to centrifugal forces, it reaches a place on the teetotum having a greater speed of rotation than the speed of the drop itself. Here the disc spins faster than the drop which seems to glide away, lagging behind the radial "spokes", as it were. That is why the drops curve, and we see the trace of curvilinear motion.

The same is true for air currents diverging from a centre of high atmospheric pressure (in "anticyclones"), or converging in a centre of low atmospheric pressure (in "cyclones"). The ink whorls depict these stupendous whirlwinds in miniature.

THE DELUDED PLANT

The centrifugal force produced by fast rotation may even outvie gravity, a point that was demonstrated by the British botanist Knight more than a hundred years ago. It is common knowledge that a young plant always directs its stem contrary to gravity, or, in plain language,

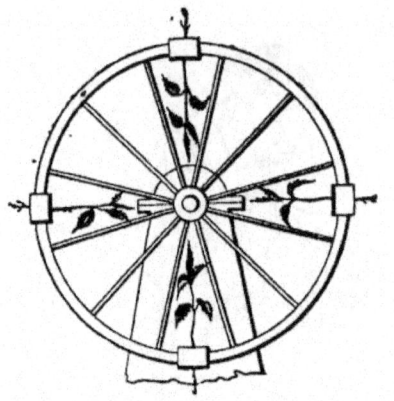

Fig. 43. Seeds germinating on the rim of a spinning wheel stem towards the axle and send their roots outwards

grows upwards. Knight, however, caused seeds to sprout inwards, from the outer rim of a quickly-spun wheel. The roots, on the other hand, were directed outwards (*Fig. 43*). He was able to fool the plant, as it were, substituting centrifugal force for gravity. The artificial gravity proved to be more powerful than the earth's natural pull—by the by, the modern theory of gravity does not present any objections, in principle, to this explanation.

"PERPETUAL MOTION" MACHINES

"Perpetual motion" is a topic that comes in for frequent mention, but I don't think all realise what it actually means. The "perpetual motion" machine is an imagined mechanism which continues its motion without end and meanwhile can also do some useful work, as lifting a load, for instance. It has never been constructed, though attempts have been made since ancient times. The futility of this task gave rise to the firm conviction that a "perpetual motion" machine is impossible, and to the law of the conservation of energy—fundamental for modern science. "Perpetual motion" as such is endless motion without any work done.

Fig. 44 depicts one of the oldest projects of a "perpetual motion" machine which certain cranks try to revive even now. Attached to the rim of the wheel are rods with weights at their ends. In any position of the wheel the weights on the right-hand side are farther from the centre than those on the left-hand side. Consequently, the right-hand weights should always outweigh the left side, thus compelling the wheel to turn. Hence the wheel should spin for ever, or at least until its axis wears through. That at any rate was what its inventor thought. Don't try to make such a machine. It will never turn. Why?

Though the right-hand weights are always farther from the centre, you are sure to have a position when they will be less in number than those on the left-hand side. Look at *Fig. 44* once again. You see only four right-hand weights and eight left-hand ones. The entire arrangement is thus balanced. The wheel will never turn; it will only swing a bit and then come to rest in this position. (The motion of this machine is explained by the so-called theorem of momenta.)

It has been proved beyond doubt that a "perpetual motion" machine as a source of energy is absolutely impossible. It is futile to undertake this task, which alchemists of yore, especially of the Middle Ages, racked their brains in vain to solve, thinking it even more tempting than the "philosopher's stone". The famous 19th-century Russian poet Pushkin describes such a dreamer, one Berthold, in his *Chivalrous Episodes*.

"'What is *perpetuum mobile*?' Martin inquired.

"'*Perpetuum mobile*,' Berthold returned, 'is perpetual motion. If I find perpetual motion I see no bounds to man's creative endeavour. For, my good Martin, while the making of gold is entrancing, a discovery perhaps, both curious and profitable, the finding of *perpetuum mobile*.... Ah, how grand that would be!'"

Hundreds of "perpetual motion" machines were invented, but none ever moved. Every inventor invariably omitted something that "upset the apple-cart".

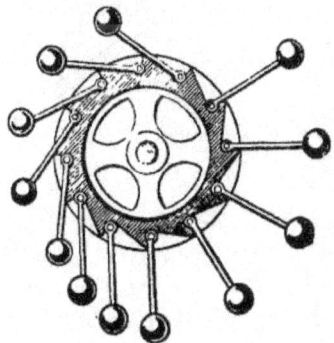

Fig. 44. An "everlastingly" moving wheel of the Middle Ages

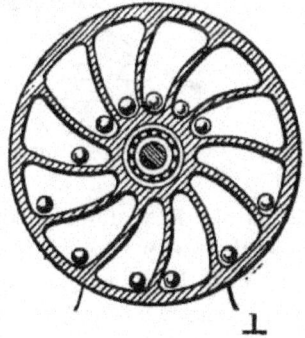

Fig. 45. A "perpetual motion" machine with balls rolling in compartments

Fig. 46. Fake *perpetuum mobile* as an advertisement
for a Los Angeles café

Fig. 45 depicts another supposed "perpetual motion" machine—a
wheel with heavy balls rolling in compartments between the outer
rim and hub. The idea was that the balls closer to the outer rim on
one side of the wheel would compel the wheel to turn by their weight.

But this will never happen—for the same reason as the wheel in *Fig.
44* doesn't turn. Still, in Los Angeles a tremendous wheel of this nature
(*Fig. 46*) was built to advertise a café. Actually it was a fake, being

turned by an artfully concealed mechanism—though people thought it was spun by the heavy balls rolling in the compartments. Other such fake "perpetual motion" machines, all set in motion by electricity, were placed in the windows of watchmaker's shops to attract the eye of the public.

Incidentally, one ad of this nature impressed my students so greatly that they wouldn't believe me when I told them that perpetual motion was impossible. Seeing is believing, they say, and when my students saw the balls rolling and turning the wheel, it seemed far more convincing than anything I could say. I told them that the fake "wonder" machine was driven by electricity from the city mains but that didn't help either. Then I recalled that on Sundays the electricity was cut off. So I advised my pupils to call on the shop on a Sunday.

"Did you see the 'perpetual motion' machine in action?" I asked afterwards.

"No," they replied, their heads ahanging, "it was covered up with a newspaper."

The law of the conservation of energy regained their confidence and they never lost faith in it again.

"THE SNAG"

Many ingenious home-taught Russian inventors tackled the fascinating problem of a "perpetual motion" machine. One, the Siberian peasant Alexander Shcheglov, is described under the name of Bürgher Prezentov by the well-known 19th-century Russian satirist Saltykov-Shchedrin in his *Modern Idyll.* Below the writer describes a visit to the inventor's workshop:

"Bürgher Prezentov was a man of some 35 summers, gaunt and pale of face. He had large pensive eyes and long hair which fell in strands onto his neck. Half of his rather roomy cottage was taken up by a big flywheel and we barely managed to squeeze in. It was a spoked wheel and had a rather large outer rim of boards nailed together like a box. Inside it was empty, and held the mechanism, the inventor's secret. There was nothing particularly cunning about it—merely bags of sand which were to balance one another. A stick in the spokes kept the wheel stationary.

"'We've heard that you've applied the law of perpetual motion in practice. Is that true?' I began.

"'I really don't know how to put it,' he returned in confusion. 'I think I've done it.'

"'Can we take a look?'

"'Pray, do! I'll be delighted.'

"He led us up to the wheel and then took us around to the other side. It was a wheel all right, from either side.

"'Does it turn?'

"'Well, it should. But it's a bit capricious.'

"'Can you take the stick out?'

"Prezentov removed it, but the wheel stood still.

"'It's up to its tricks again!' he repeated. 'It needs an impetus.'

"He gripped the rim with both hands, swung it back and forth several times, then pushed it with all his might. The wheel began to turn. It made several turns rather quickly and smoothly. One could hear the bags of sand inside the rim banging against the boards and sliding away. Then the wheel began to turn more and more slowly. We heard a rasping and a creaking and, finally, the wheel stopped altogether.

"'Must be a snag somewhere,' the inventor explained in confusion as he strained and swung the wheel again. But the result was the same.

"'Perhaps you forgot friction?'

"'I didn't.... Friction you say? It's not because of that. Friction's nothing. Sometimes it makes you happy and then, bang, it's up to its tricks, gets ornery, and that's that. If the wheel were made of real stuff, not scraps!'"

It was of course not the "snag" or the "real stuff" that was at fault, but the wrong principle at the root. The wheel turned for a time owing to the impetus that the inventor gave it, but was bound to stop when friction exhausted the imparted outside energy.

"IT'S THEM BALLS THAT DO IT"

The writer Karonin (the pen-name of N. Y. Petropavlovsky) describes another Russian "perpetual motion" machine inventor in his story "*Perpetuum Mobile*". This was Lavrenti Goldyrev, a peasant from

Perm Gubernia who died in 1884. Karonin, who changed the name in' the story to Pykhtin, describes the machine in great detail.

"Before us was a large queer machine resembling at first glance the sort of thing a blacksmith uses to shoe horses on. We could see some badly planed wooden pillars and beams and a whole system of flywheels and gear wheels. It was all a very clumsy-looking affair, rough and ugly. Several iron balls lay on the floor underneath the machine and there was a whole pile of them a bit to the side.

"'Is that it?' the major-domo asked.

"'That's it.'

"'Well, does it turn?'

"'How else?'

"'Have you got a horse to turn it?'

"'A horse? What for? It turns by itself,' Pykhtin returned and began to demonstrate the monster's workings.

"The main role was played by iron balls heaped up nearby.

"'It's them balls that do it. Look. First it goes whack into this scoop. Then it flies like lightning along that groove, is scooped up by that scoop, flies like mad back to that wheel and again gives it a good push so hard that it even begins to whine. Meanwhile another ball is on its way. Again it flies along and goes whack here. From here it dashes along the groove and strikes that scoop, skips to the wheel, and again whack! That's how it goes. Wait, I'll start it off.'

"Pykhtin darted to and fro, hastily collecting the scattered balls. Finally, after heaping them up into a pile by his feet, he picked one up and threw it with all his might at the nearest scoop on the wheel. Then he quickly picked up a second, then a third. The noise was something unimaginable. The balls clanked against the iron scoops, the wheel creaked, the pillars groaned. An infernal whine and racket filled this gloomy place."

Karonin claims that Goldyrev's machine moved. But this was patently a misunderstanding. The wheel could have turned only while the balls were dropping down—at the expense of the potential energy accumulated when lifted, much in the manner of the weights of a pendulum clock. However, it couldn't have turned long because when all the lifted balls had "whacked" against the scoops and had slipped down, it would stop—provided it hadn't stopped before by the counter-effect of all the balls it was supposed to lift.

Later on, Goldyrev became disappointed in his invention when at an exhibition in Yekaterinburg, where he showed it, he saw real industrial machines. When asked about his "perpetual motion" contraption, he dejectedly replied: "The devil take it! Tell 'em to chop it up for firewood."

UFIMTSEV'S ACCUMULATOR

Ufimtsev's so-called accumulator of kinetic energy well illustrated the pitfalls that may trap a cursory observer of a "perpetual motion" machine. Ufimtsev, an inventor from Kursk, devised a new kind of windmill power station with a cheap flywheel type of "inertia accumulator". In 1920 he built a model of it, shaped as a disc that spun round a vertical axis set on ball bearings inside an air-free jacket. When revved to 20,000 r.p.m., the disc was able to turn for 15 days on end. The unthinking observer could well believe that he had before him a real "perpetual motion" machine.

"A MIRACLE, YET NOT A MIRACLE"

The futile search for a "perpetual motion" machine clouded many lives. I once knew a factory worker who sank into absolute destitution, spending all his earnings and savings in the delusion that he could make a "perpetual motion" machine. Poorly clad and always hungry, he would beg everyone he met to give him some money to make the "finished model", which would "certainly move". It was a great pity to see this man suffering so much only because of his ignorance of the rudiments of physics.

It is curious to note that whereas the search for a "perpetual motion" machine was always abortive, the profound realisation of its impossibility, on the contrary, often led to discoveries of great value.

A wonderful illustration in point is the method which the remarkable Dutch scientist Stevin, who lived at the turn of the 16th century, evolved to establish the law of the equilibrium of forces on an inclined plane. He deserves far greater fame than befell him for his many major discoveries that we now constantly address ourselves to. These are decimal fractions, the introduction of denominators in algebra, and the establishment of the hydrostatic law that Pascal rediscovered later.

Stevin evolved the law of the equilibrium of forces on an inclined plane without invoking the rule of the parallelogram of forces. He proved it with the aid of a drawing, which is reproduced in *Fig. 47*. A chain of fourteen identical spheroids is slipped round a three-sided prism. What happens to it? The bottom, which droops garland-like, is in a state of balance, as you see. But do the other two [parts balance each other? In other words, do the two spheroids on the right offset the four on the left? The answer is yes. Otherwise the chain would keep on rolling of its own accord from right to left for ever. Otherwise other spheroids

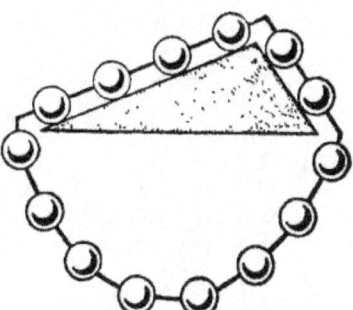

Fig. 47. "A miracle, yet not a miracle"

take the place of those that slide ⌐off and ⌐equilibrium would never be restored. But we know that a chain disposed in this fashion does not move of its own accord at all It is quite obvious that the two spheroids on the right really offset the four on the left.

It seems a minor miracle, doesn't it? Two spheroids pull with the same force as four! This enabled Stevin to deduce an important law of mechanics. This is how he reasoned. The two parts—the long one and the short one—possess a different weight, one being as many times heavier than the other as the longer side of the prism is longer than the short side. Consequently, any two linked loads in general balance on tilted planes, provided their weight is directly proportional to the length of these planes.

When the short plane is vertical we get a well-known law of mechanics, which is: to hold a body in place on a tilted plane we must act in the direction of this plane with a force as many times less the weight of the body as the length of the plane is greater than its height. So did the idea that a "perpetual motion" machine is impossible led to an important discovery in the realm of mechanics.

MORE "PERPETUAL MOTION" MACHINES

Fig. 48 shows a heavy chain fitted around wheels in such a way that the right-hand part is always longer than the left-hand part, whatever its position. The inventor thought that since the right-hand part would always weigh more than the left-hand part, it would always outweigh the left-hand part and thus cause the entire arrangement to keep going. But does this really happen? Of course not. You already know that the heavier part of a chain may be offset by the lighter part, provided they are pulled by forces acting at different angles. In this particular system, the left-hand part of the chain droops vertically down, while the right-hand part is inclined. So, though it is heavier, still it cannot pull over the left-hand part and we do not achieve the "perpetual motion" expected.

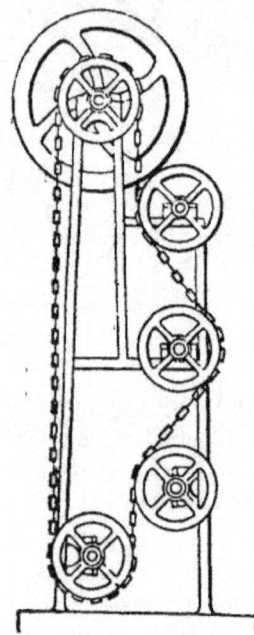

Fig. 48. Is this a "perpetual motion" machine?

I think the cleverest "perpetual motion" machine ever invented was one displayed at the Paris Exposition in the 1860's. It consisted of a large wheel with balls rolling about in its compartments. The inventor claimed that nobody would ever be able to stop the wheel. Many visitors tried to stop it but it went on turning as soon as they took their hands off it. Not a single person realised that the wheel turned precisely because of the effort he made to stop it. The backward push he gave to stop it wound up the spring of an artfully concealed mechanism.

THE "PERPETUAL MOTION" MACHINE PETER
THE GREAT WANTED TO BUY

Preserved in archives is a bulky correspondence which Peter the Great of Russia carried on between 1715 and 1722, when he wanted to buy a "perpetual motion" machine that had been devised in Germany by one Councillor Orffyreus. This man whose "self-moving wheel" won him nation-wide fame consented to sell it to the tsar only for a princely sum. Peter the Great's librarian Schumacher, whom the tsar had sent to Western Europe to collect rare oddities, reported the following, when asked to negotiate the purchase:

"The inventor's last words were: One hundred thousand thalers and you get the machine."

As for the machine itself, according to Schumacher, the inventor claimed that it was no fake and that it could not be defamed "except out of malice, and the whole world is full of spiteful people whom one cannot believe".

In January 1725 Peter the Great decided to go to Germany to see this notorious "perpetual motion" machine himself, but he died before he could accomplish his purpose.

Who was this mysterious Councillor Orffyreus and what was his "famous machine" really like? I was able to learn something both about the Councillor himself and his machine.

Orffyreus's real name was Bessler. He was born in Germany in 1680. He studied theology, medicine and painting before he essayed the "perpetual motion" machine. Among the many thousands who tried to invent such a machine he is probably the most famous and, at any rate, the luckiest. Till the end of his days—he died in 1745—he lived in comfort on the income he netted by demonstrating his contraption.

Fig. 49 is a reproduction of a drawing from an old book depicting Orffyreus's machine as seen in 1714. It shows a large wheel which apparently not only turned by itself, but even lifted a heavy load to quite a height.

The fame of this "miracle" machine, which the learned councillor first exhibited at various market fairs, quickly spread throughout Germany. Soon Orffyreus acquired powerful patrons. The Polish

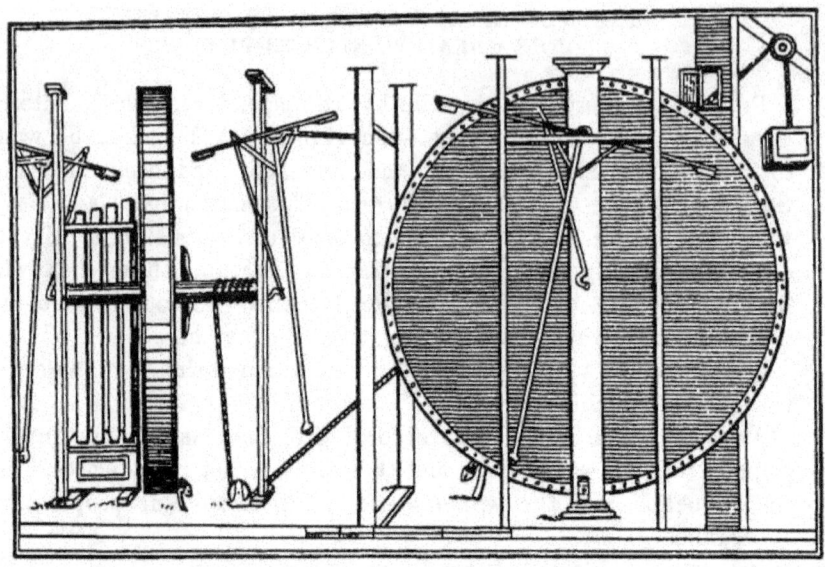

Fig. 49. Orffyreus's self-moving wheel which Peter the Great wanted to buy. (From an old drawing.)

king displayed interest and then the Landgrave of Hesse-Cassel patronised the inventor, placing his castle at the latter's disposal and subjecting the machine to every kind of trial.

On November 12, 1717, the machine was placed in a room all apart and set into motion. The room was then locked and sealed, and two grenadiers were posted outside. For a whole fortnight, until the seal was broken on November 26, no one dared to come near. Then the room was unlocked and the Landgrave and his retinue entered. The wheel was still spinning "with undiminishing speed". It was stopped, inspected carefully, and again set going. Now the room was locked and sealed for 40 days on end with grenadiers again stationed at the door. The seal was broken on January 4, 1718. A commission of experts entered and found that the wheel was still going. But this did not satisfy the Landgrave and he staged a third trial, locking up the machine for two whole months at a stretch. When he found the wheel still going even after that, he was delighted. He granted the inventor a parchment to certify that his "perpetual motion" machine did 50 revolutions per minute, could lift 16 kg to the height of 1.5 m and could also work a grinder and bellows. With this document in his pouch, Orffyreus travelled the length and breadth of Europe. He apparently netted a princely income, considering that he consented to sell his machine to Peter the Great for not less than 100,000 rubles.

The fame of the councillor's marvel quickly spread, finally reaching the ears of Peter the Great, who had a very weak spot in his heart for all sorts of curious and cunning artifices, and, naturally, it intrigued him greatly. His attention had been called to it back in 1715 when travelling abroad, and it was then that he charged the celebrated diplomat A. I. Ostermann to inspect it. The latter soon forwarded an extensive report about the machine though he had not been able to see it with his own eyes. The tsar even thought of inviting Orffyreus as an eminent inventor to his court to take up service and asked the then well-known philosopher Christian Wolf to give his opinion.

Orffyreus was showered with offers, one better than the other. Kings and princes bestowed munificent awards. Poets composed odes in honour of his wonder-wheel. But there were some who thought him a charlatan. The more daring openly accused him, even offering 1,000 marks to anyone who would come forth and expose the councillor. One lampoon against him gave a drawing which is reproduced in *Fig. 50* and which provides a rather simple explanation for the mystery—a cunningly hidden person who pulled at a rope wound round that part of the axle which was concealed in the pillars supporting the wheel.

The trick was bared by chance only because the councillor had had a tiff with his wife and maid who had both been initiated into the secret. Otherwise we would probably still be guessing. It seemed that the notorious machine was indeed turned by a hidden person—Orffyreus's brother, or maid—pulling at a slender cord. But the councillor did not lose face, persistently assuring all and sundry even on his deathbed that his wife and maid had maligned him out of spite. However, trust in him was shattered. No wonder he tried to drum into the head of the tsar's envoy, Schumacher, the point that human beings were full of malice.

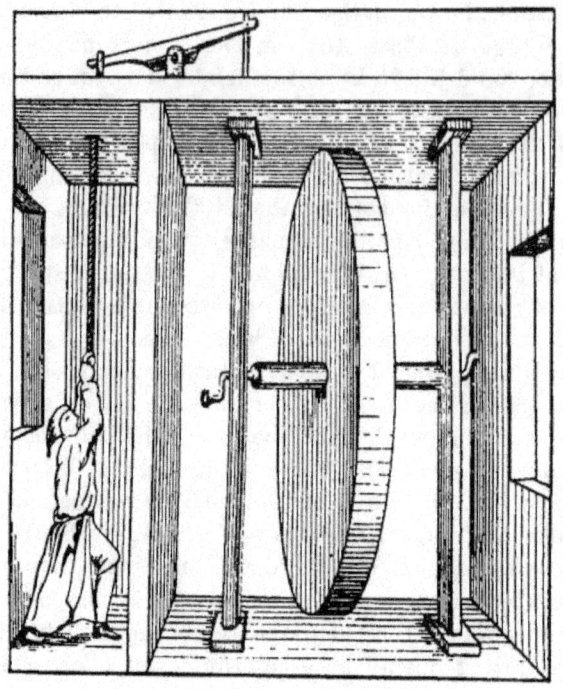

Fig. 50. The secret of Orffyreus's machine.
(From an old drawing.)

Around the same time there also lived in Germany another renowned
"perpetual motion" machine inventor, one Hertner. Schumacher wrote
of his contraption the following: "Herr Hertner's *perpetuum mobile*,
which I saw in Dresden, consists of tarpaulin filled with sand and a
grinder-like machine which turns forwards and backwards by itself.
However the inventor says it cannot be made larger." Undoubtedly
this machine, too, gave no "perpetual motion", being at best an artfully
contrived device with a just as artfully concealed—living—but by
no means "perpetual motion" machine. Schumacher was right when
he wrote to Peter the Great that French and English scholars "mock
these *perpetuum mobiles* as objectionable to principles of mathematics".

CHAPTER FIVE

PROPERTIES OF LIQUIDS AND GASES

THE TWO COFFEE-POTS

Fig. 51 shows two coffee-pots of the same width. One, however, is taller than the other. Which of the two will hold more? An unthinking person would probably point to the taller one. However, we would be able to fill it up only to the level of its spout, and if we poured more in, it would all spill out Now since the spouts of both coffee-pots are on the same level, the lower one takes just as much liquid as the taller one does. You will easily realise why. The coffee-pot and its spout are two communicating vessels and hence inside both the liquid should be at an identical level, even though the liquid in the spout weighs much less than that in the coffee-pot proper. Unless the spout is high

Fig. 51. Which coffee-pot takes more?

enough, you will never be able to fill the coffee-pot up to the top; the water will simply keep on spilling out. Usually the spout is even a bit higher than the top of the coffee-pot to enable one to incline it without spilling out its contents.

IGNORANCE OF ANCIENTS

Romans today still use what is left of the aqueducts that their forefathers built. Though the Roman slaves of old did a very good job, we can't say that of the Roman engineers in charge. Their knowledge of elementary physics was plainly inadequate. *Fig. 52* reproduces a picture preserved at the German Museum in Munich. As you see, the Romans did not sink their water systems in the ground but placed

Fig. 52. The aqueducts of ancient Rome

them on high supports of masonry. Why? Aren't underground pipes of the type we use today simpler? Roman engineers of old had a very hazy notion, however, of the laws of communicating vessels. They feared that in two reservoirs connected by a very long pipe, the water would not rise to the same level. Furthermore, if the pipes were laid in the ground and followed the natural relief, in some places the water would have to flow upwards, and this was something the Romans were afraid it would not do. That is why their aqueducts usually slope all along the way. They often had either to take the pipes on a roundabout route or erect tall arches. One Roman aqueduct, known as the Aqua Marcia, is 100 km long, though it is half the distance between its two points as the crow flies. As you see, the ancient Romans' ignorance of an elementary law of physics caused 50 km of extra masonry to be built.

LIQUIDS PRESS ... UPWARDS

Even people who have never studied physics know that liquids press down on the bottom of the vessels holding them and sideways at the walls. Many, however, have never suspected that liquids also press upwards. An ordinary lamp-glass will easily reveal this. Cut out of a piece of thick cardboard a disc large enough to cover the top of the lamp-glass. Cover the top of the glass with it and then dip the glass into a jar of water as shown in *Fig. 53*. To prevent the disc from slipping off when the lamp is immersed, tie a piece of thread to it and hold it as shown, or simply press it down with your finger. After you have dipped the glass far enough, you can let the thread, or your finger, go. The disc will remain where it is, being kept in place by the water pressing up on it.

If you want to, you can even gauge the value of this upward pressure. Carefully pour some water into the glass. As soon as the level of the water in the glass reaches that of the water in the jar, the disc slips

Fig.53. A simple way to demonstrate that liquids |press upwards

off, because the pressure exerted by the water on the disc from below is offset by the pressure exerted on it from above by the column of water in the glass, the height of which is equal to the depth to which the glass has been dipped. Such is the law concerning the pressure that a liquid exerts on any immersed body. This incidentally results in that "loss" of weight in liquids of which Archimedes's famous principle speaks.

With the help of several lamp-glasses of different shapes but with tops of one and the same size you may test another law dealing with liquids: that the pressure a liquid exerts on the bottom of the containing vessel depends only on the size of the bottom and the height of the "column" of liquid; it does not depend at all on the vessel's shape. This is how you test this law. Take different glasses and dip them to one and the same depth. To see that no mistakes occur, first glue strips of paper to the glasses at equal heights from the bottom. The cardboard disc you used in the first experiment will slip off every time you pour in water to the same level (*Fig. 54*). Consequently the pressure exerted by columns of water of different shapes is the same as long as the bottom and height are the same. Note that it is the *height*, and not the *length*, that is important, because a long but *inclined* column exerts exactly the same

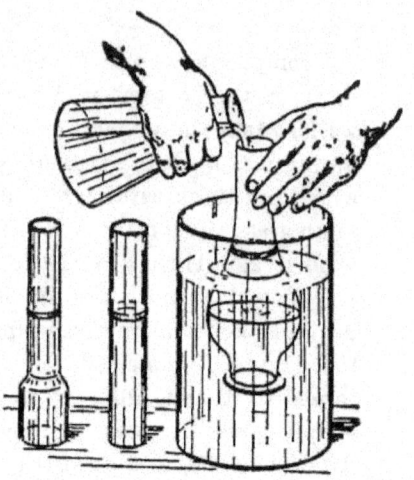

Fig. 54. The pressure liquid exerts on the bottom of the vessel depends only on the area of the base and the liquid's height. The drawing shows you how to check this

pressure on the bottom as is exerted by a shorter but perpendicular column as *high* as the inclined one—provided, of course, the bottom of each is the same.

WHICH IS HEAVIER?

Fig. 55. Both pails are full to the rim. One has a piece of wood in it. Which is heavier?

Place a pail of water, full up to the rim, on one pan of a pair of scales. Then put on the other pan another pail of water, *also full up to the rim*, but with a piece of wood floating in it (*Fig. 55*). Which of the two is heavier? I asked this of different people and got contradictory answers. Some said the pail with the piece of wood in it would be heavier because it held a piece of wood in addition to the water. Others said the pail of water without the piece of wood would be heavier, since water generally weighs more than wood. Neither were right. Both pails *weigh the same.* The second pail, true, contains less water than the first one, because the wood displaces some of the water. But, according to the related law, every *floating* body displaces with its immersed part exactly *as much liquid* (in weight) *as the whole of this body weighs.* That is why the scales balance.

Now try to solve another problem. Take a glass of water, put it on one of the pans, and put a weight next to it. Balance the scales. Then drop the weight next to the glass into it. What happens to the scales? According to Archimedes's principle, in the water the weight should weigh less than when on the pan.

Consequently, oughtn't this pan rise? However, the pans maintain their equilibrium. Why? When dropped into the glass the weight displaced some of the water which then rose to a level higher than before. This added to the pressure exerted on the bottom of the vessel, which thus sustained an additional force equivalent to the weight lost by the weight.

A LIQUID'S NATURAL SHAPE

We are used to thinking that liquids have no shape of *their own.* That is not true.

The natural shape of any liquid is that of a sphere. As a rule, gravity prevents liquids from assuming this shape. A liquid either spreads in a thin layer if spilled out of a vessel, or takes the vessel's shape. But when inclosed in another liquid of the same specific weight, it, according to Archimedes's principle, "loses" its weight, seeming to weigh nothing; now gravity has no effect on it and it assumes its natural spherical shape.

Since olive oil floats in water but sinks in alcohol we can mix the two in such proportions that the oil will neither sink nor float in this mixture. An odd thing happens when we drip in a little oil with the help of an eyedropper. The oil collects into a large round drop which neither floats nor sinks, but hangs suspended (*Fig. 56*). To get a true image of the sphere, you should do the experiment in a flat-walled vessel—or in one of any shape but placed inside a flat-walled vessel full of water.

You must do this experiment patiently and carefully, because otherwise you will get several smaller drops instead of a large one. Don't feel disheartened if it doesn't work out; even then it's sufficiently illuminating.

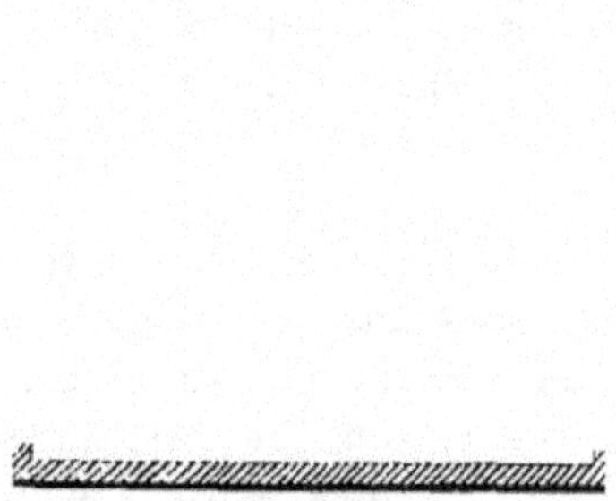

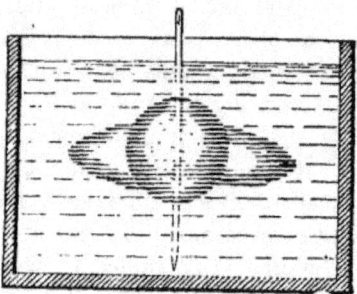

Fig. 56. Oil inside diluted alcohol collects into a drop which neither sinks nor floats. (Plateau's experiment.)

Fig. 57. A ring is given off when the oil drop in the alcohol is spun by means of a rod

Let's carry this experiment further. Take a long stick or a piece of wire and transfix the oil drop. Start turning. The drop also participates in this revolution. You get still better results by attaching to the stick or wire a small cardboard disc soaked in oil and inserting it fully in the drop you are twirling. The spin compels the drop to compress and then give off a ring a few seconds later (*Fig. 57*). As it breaks up the ring creates new drops which continue to revolve round the central one.

The Belgian physicist Plateau was the first to conduct this instructive experiment, of which I have given you the classical description. It would be much easier—and just as instructive—to do this experiment in another way. Take a small tumbler, rinse it with water, and fill it with olive oil. Place it on the bottom of a larger glass. Then carefully pour into the glass enough alcohol to cover the tumbler. Gradually add a little water with the help of a spoon. Do this very carefully, so that the water drips down the walls of the glass. The top of the oil in the tumbler starts to bulge, and when enough water has been poured in, the oil rises up from the tumbler in a rather large drop to hang suspended in this mixture of alcohol and water (*Fig. 58*).

For want of alcohol you can use aniline instead. Aniline is a liquid which is heavier than water at room temperature but lighter than water when heated to 75-85 °C. By heating up the water, we can make the aniline

Fig. 58. Plateau's experiment simplified

swim inside it and assume the form of a large drop. At room temperature you can suspend an aniline drop in a solution of table salt. Another convenient liquid is the dark-crimson orthotoluidine, which at 24 °C has the same density as salt water, into which it is poured.

WHY IS SHOT ROUND?

I noted earlier that any liquid will assume its natural spherical shape when gravity ceases to act on it. You need only remember what I said before about a falling body having no weight and discount the negligible atmospheric resistance when a body starts to fall (raindrops accelerate only when they start to fall; by the second half of the first second the fall already becomes *uniform* and the drop's weight is offset by atmospheric resistance which grows together with the velocity of the falling drop) to realise that falling portions of liquid should also take on a spherical form.

That is really so. Falling raindrops are indeed round in shape. Shot is nothing but solidified drops of molten lead which in the process of making are dropped from a great height into a cold water bath where they solidify in the shape of absolutely right spheres. Shot is also called "tower" shot because in its making it is dropped from the top of a tall "shot tower" (*Fig. 59*). These towers are metal structures 45 m high. At the top they have a shot-pouring shop with boilers for melting the lead, and at the bottom—a water bath. The ready shot is then graded and processed. The drop of molten lead solidifies into shot while falling. The water bath is needed merely to soften the impact and to prevent the shot from losing its spherical shape (shot with a diameter of more than 6 mm, so-called canister shot, is made differently, by chopping off pieces of wire, which are then rolled into balls).

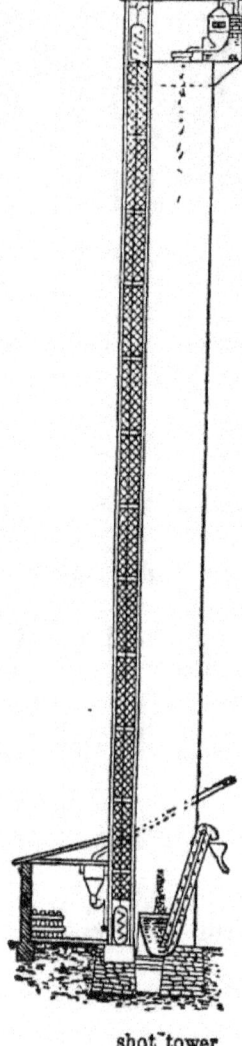

shot tower

THE "BOTTOMLESS" WINEGLASS

Fill a wineglass with water right up to the very rim. Take some pins. Do you think place could be found in the wineglass for a couple of them? Try and see.

Throw the pins in and count them as you do. But be careful about it. Take the pin by its head and dip its point into the water. Then carefully let go, without pushing it or exerting any pressure, so that water is not spilled out. As you drop the pins in, they fall to the bottom, but the level of the water is the same. You drop in ten, then another ten, and then another ten. The water does not spill out. You can go on till there are a hundred at the bottom of the glass. But still no water has spilled out (*Fig. 60*). Nor, for that matter, has it risen to any noticeable degree above the rim.

Add some more pins. Now you can even count them in hundreds. You may have as many as 400 pins in the glass, but still no water spills out. However, now you see that the surface is bulging above the rim. Therein lies the answer to this so far incomprehensible phenomenon. Water scarcely wettens glass as long as it is a little greased, and the rim of the wineglass—like all the chinaware and glassware we use for that matter—is sure to have some traces of grease which are left when we touch it with our fingers. And as it doesn't wetten the rim the water displaced by the pins bulges. You can't see it, but if you went to the pains of reckoning the volume of one pin and of comparing it with the volume of the bulge above the rim of the wineglass you would realise that the former volume is hundreds of times smaller than the latter, which explains why a "full" wineglass will still have enough room for another few hundred pins.

Fig. 60. How many pins in the wineglass?

The wider the mouth of the wineglass is, the more pins it can take, because there is a larger bulge. A rough reckoning will make the point clear. A pin is about 25 mm long and half a millimetre thick. You can easily reckon the volume of this cylinder by invoking the well-known geometrical formula $\left(\frac{\pi d^2 h}{4}\right)$; it will be equal to 5 mm^3. Together with the head, the pin will have a total volume of not more than 5.5 mm^3. Let us now reckon the volume of the water in the bulge. The diameter of the wineglass mouth is 9 cm, or 90 mm. The area of such a circle is about 6,400 mm^2. Assuming that the bulge is not more than 1 mm high, we thus get a volume of 6,400 mm^3, which is 1,200 times more than the volume of the pin. In other words, a "full" wineglass of water can take more than a thousand pins. And indeed we can get the wineglass to take a thousand pins if we are careful enough. To the eye they seem to occupy the whole of the wineglass and even stick out of it. But still no water spills out.

UNPLEASANT PROPERTY

Anyone who has ever had to handle a kerosene lamp most likely knows what annoying surprises it can spring on one. You fill a tank with it and then wipe the tank dry on the outside. An hour later it's wet again. You have only yourself to blame. You probably didn't screw on the burner tight enough, and the kerosene, trying to spread along the glass, seeped out. To avert such "surprises", screw the burner on as tight as you can. But when you do that, don't forget to see that the tank is not full up to the brim. When it warms up, kerosene expands rather considerably—increasing in volume by a tenth every time the temperature rises by another 100°. So if you don't want the tank to explode, you must leave some room for the kerosene to expand.

The property of kerosene to seep through causes unpleasant things aboard ships whose engines burn kerosene or oil. When due precautions are not taken, it is absolutely impossible to use such ships to carry any other cargoes except kerosene or oil, because when they seep out through unnoticeable crevices in the tanks these liquids spread not only to the metal surfaces of the tanks but literally everywhere, even to the clothing of the passengers to which they impart a smell that nothing will kill.

Attempts to fight this evil are often to no avail. Jerom K. Jerome, the British humorist, wasn't guilty of much of an exaggeration when in his *Three Men in a Boat* he wrote of paraffin oil, which is remarkably alike kerosene.

"I never saw such a thing as paraffin oil is to ooze. We kept it in the nose of the boat, and, from there, it oozed down to the rudder, impregnating the whole boat and everything in it on its way, and it oozed over the river, and saturated the scenery and spoilt the atmosphere. Sometimes a westerly oil wind blew, and at other times an easterly oil wind, and sometimes it blew a northerly oil wind, and maybe a southerly oil wind; but whether it came from the Arctic snows, or was raised in the waste of the desert sands, it came alike to us laden with the fragrance of paraffin oil.

"And that oil oozed up and ruined the sunset; and as for the moonbeams, they positively reeked of paraffin....

"We left the boat by the bridge, and took a walk through the town to escape it, but it followed us. The whole town was full of oil." (Actually it was only the clothing of the travellers that reeked of paraffin oil.)

The property kerosene has of wettening the outer surface of tanks led people to wrongly think that kerosene could ooze through metal and glass.

THE UNSINKABLE COIN

It's to be found not only in fairy tales. A few easy experiments will show you that such things really exist. Start with a small object—a needle, for instance. It seems impossible to make a steel needle float, doesn't it? But it isn't really so hard to do. Place a strip of cigarette paper on top of the water in a glass and an absolutely dry needle on top of the paper. Carefully remove the cigarette paper in the following

way. Take another needle or a pin and, gradually working to the middle, gently push the strip of paper into the water. When the strip is soaked through, it will sink, but the needle will continue to float (*Fig. 61*). By moving a magnet at water level from outside the glass you can even make the floating needle spin round.

With a little experience, you can dispense with the cigarette paper entirely. All you need do is to take the needle by the middle and, holding it parallel to the water, drop it from a small height. You can make a pin, which like the needle must not be thicker than 2 mm, a light button, or some small metal object float in the same way. When you have got the knack of it, try a coin.

All these metal objects float because water hardly wettens metal covered with a very thin film of grease from our hands. You can even see the depression a floating needle makes on the surface of the water. Trying to regain its original position, the surface film buoys up the needle which is

Fig. 61. A floating needle. Top: a cross-section of the needle (2 mm thick) and the depression it makes (a twofold magnification). Bottom: how to make the needle float by using a strip of paper

also buoyed up by a force equal to the weight of the water displaced by the needle. The easiest way of making a needle float is, of course, to cover it with grease. Then it will never sink.

CARRYING WATER IN A SIEVE

Neither is this something that can be done only in a fairy tale. Physics can help us to undertake this seemingly impossible task. Take a wire sieve of 15 cm across with holes not smaller than 1 mm in diameter and dip it into melted paraffin, to cover it with a thin, barely discernible film.

Your sieve remains a sieve; it still has holes in it through which a pin can go quite freely, but now you can carry water in it—even quite a lot of it. Only be careful when pouring the water in and see to it that you don't jolt the sieve while doing that.

Why doesn't the water drip through? Failing to wetten the paraffin, the water forms a thin film which bulges through the holes of the sieve; it is this film that keeps the water from dripping through (*Fig. 62*). This waxed sieve will even float, which means that you can not only carry water in a sieve, but also use it as a boat.

This seemingly paradoxical experiment explains several ordinary things to which we are too accustomed to ever think of why they are done. The tarring of barrels and boats, the greasing of corks and stoppers with fat, the painting of roofs with oil paint and, generally, the coating with oily substances of everything we want to make impervious to water, as well as the rubberising of cloth, is the same as making the sieve we just described, with the exception that the sieve, of course, seems exceedingly unusual.

Fig. 62. Why the sieve carries water

FOAM HELPS ENGINEERS

The experiment of the floating steel needle or copper coin bears some resemblance to a process employed in mining to "enrich" ores, i.e., to increase the content of the minerals in them. Engineers know many methods for dressing ores, but the one we have in mind and which is called "flotation" is best; it is successful when all other methods fail.

Flotation consists in the following. Finely ground ore is loaded into a bath containing water and oily substances that inclose the mineral particles in a very thin film which water cannot wet. Air is then blown in to form a foam composed of a multitude of tiny bubbles. The greased particles of the mineral attach themselves to the air bubbles and rise up with them much in the same way as an air balloon lifts a gondola (*Fig. 63*). The particles of ore gangue that have no grease envelope cannot attach themselves to the air bubbles and sink. Note that the air bubbles in the foam are much bigger than the particles they carry and are well able to lift the solid speck up. As a result, nearly all the particles of

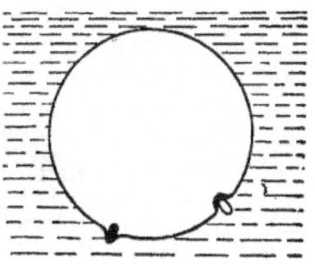

Fig. 63. The essence of flotation

the mineral are floated on top in the foam which is skimmed off for further processing, during which the so-called concentrated ore—which is dozens of times richer in content than the original ore—is separated. Flotation techniques are so well elaborated that a judicious choice of reagents will separate the mineral from the ore gangue in every particular case.

Incidentally, we have a chance accident, and no theory, to thank for the flotation method. One day, at the end of the past century, Carrie Everson, an American schoolmistress, was washing greasy sacks that had been used to stack copper pyrites. She happened to notice that the pyrite particles left in the sacks floated together with the lather. It was this that suggested the flotation method.

FAKE "PERPETUAL MOTION" MACHINE

You will sometimes find the following contraption (*Fig. 64*) described as a genuine "perpetual motion" machine. Oil (or water) poured into a vessel is soaked up by wicks at first into one vessel and then by more wicks into another vessel still higher up. The top vessel has a grooved outlet through which the oil pours onto a paddle wheel, causing it to turn. From the bottom tank the oil is again soaked up by wicks to the top. Thus, the oil supposedly never stops pouring onto the paddle wheel, making the wheel turn for ever and ever.

Fig. 64. Non-existent "perpetual motion" machine

If the people who described this thing were to take the pains to make it, they would realise that not a single drop of oil would ever reach the upper vessel, let alone make the wheel go. Incidentally, we don't necessarily have to make this contraption to realise that this is so. Indeed, why should the inventor think the oil should necessarily flow off the upper bent portion of the wick? It is quite true that capillary forces, having overcome gravity, lift the oil up the wick. But it is these same forces that prevent the oil in the pores of the soaked wick from oozing off. Even supposing for a moment that the oil will reach the upper vessel of our fake "perpetual motion" machine due to capillary forces, we shall have to admit that the same wicks which supposedly lift the oil up would themselves lower it to the bottom tank.

The contraption we have just mentioned resembles another water-driven one, invented by the Italian mechanic Strada the Elder way back in 1575. *Fig. 65* shows you this amusing device. As it turns, an Archimedes's screw lifts water to the upper tank, from which it pours out through a groove to strike at the paddles of the tank-filling wheel

Fig. 65. An ancient design of a water-driven "perpetual motion"
machine to turn a grinding stone

shown in the bottom right-hand corner. This wheel turns a grinder and simultaneously operates by means of several gears the same Archimedes's screw which lifts the water to the upper tank. To make a long story short, the screw turns the wheel and the wheel turns the

91

screw! If such contraptions were possible, the simplest thing would be to throw a rope over a pulley and tie identical weights to each end. As one weight fell it would lift the other one, which, dropping in turn, would lift the first one. Wouldn't that be a fine "perpetual motion" machine?

BLOWING SOAP BUBBLES

Do you know how to blow soap bubbles? It is not so simple as it seems. I, too, thought there was nothing particular in it until I saw for myself that the ability to blow big beautiful bubbles is in its way an art that needs some experience. But is it really worth while doing such a seemingly silly thing as blowing soap bubbles? After all, they have won a rather bad reputation among the laymen. Physicists have other views, however. "Blow a soap bubble," said the great British physicist Kelvin, "and observe it. You may study it all your life, and draw one lesson after another in physics from it."

Indeed, that magic iridescence on the slimmest of soap films enables the physicist to gauge the length of light waves, while a study of the tension of these gossamer films helps him to formulate the laws governing the interaction of forces between particles—those self-same forces of cohesion without which the world would be but a cloud of the finest dust.

The few experiments described below do not have such serious aims; they are given simply to provide instructive entertainment and to teach you how to blow soap bubbles. In his book *Soap Bubbles and the Forces Which Mould Them*, the British physicist Charles Boys describes at length many different experiments that one can stage with these bubbles. So if you are interested in them, let me refer you to this wonderful book.

Below you will find only a few of the simplest experiments. Ordinary laundry soap will do—toilet soaps are less suitable for the purpose. But you can also use pure olive-oil or almond-oil soap, which is best for obtaining large and beautiful bubbles. Carefully dissolve a cake of soap in pure cold water till you get a rather thick lather. Pure rain water or melted snow is best but you may use cooled boiled water instead. To prolong the life of the bubbles Plateau suggests adding glycerin to the lather in a mixture of one part to every three. Skim the

92

froth and the small bubbles off with a spoon and then dip in the lather a slender clay pipe, with its end preliminarily soaped both on the inside and outside. Good results can be achieved also by using straws of about 10 cm long, that are split at the bottom in the form of a cross.

This is how you blow the bubble. Dip the pipe into the lather, holding it vertically so that it becomes covered with film. Then gently blow at the other end. As the bubble is filled with warm air from our lungs—which is lighter than the air in the room—it will float up at once as long as you can blow a bubble of some 10 cm across; otherwise you must add more soap until you can blow bubbles of this diameter. This alone is not enough; there is another test that you must make. After you blow the bubble, dip your finger in the lather and try to pierce the bubble with it. If it doesn't burst you can start experimenting. If it does—add a little more soap. Do the experiments slowly, with care, and without undue haste. The room must be well lit; otherwise the proper iridescence will be lacking. Now for a few entertaining experiments.

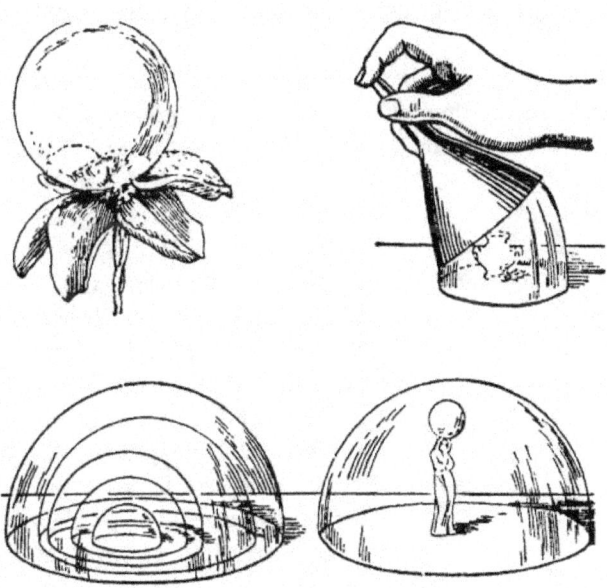

Fig. 66. Soap bubbles

1) *A flower in a bubble.* Pour the lather three millimetres deep into a plate or tray. Then place a flower or a little vase in the middle and cover it with a glass funnel. Slowly lift the funnel, blowing meanwhile in its narrow end to get a soap bubble. When the bubble is large enough, tilt the funnel as shown in *Fig. 66* and release the bubble. Your flower or vase will be under a transparent, semicircular, iridescent soap bubble. You can take a *statuette* instead of a flower and crown it with a small soap bubble as shown in *Fig. 66*. To get the smaller bubble, you must spill a little lather on top of the statuette before you blow the big bubble. Then pierce the big bubble with a pipe and blow out the small bubble inside.

2) *A nest of bubbles (Fig. 66).* Take the funnel you used for the previous experiment and blow a large bubble as you did before. Then take a straw and dip it into the lather, leaving only the very end, which you blow through, dry. Gently pierce the wall of the first bubble till you get to the middle. Then slowly draw the straw back without bringing it out, and blow out a second bubble inside the first. Repeat to get a third bubble inside the second, a fourth inside the third, and so on.

3) *A cylindrical bubble (Fig. 67).* For this purpose you must have two wire rings. Blow an ordinary round bubble onto one of them, the lower one. Then take the second ring, wet it and attach it to the top

Fig. 67. How to make a cylindrical soap bubble

of the bubble. Lift it until the bubble assumes a cylindrical shape. Note that if you lift the upper ring to a height more than the ring's circumference, half of the cylinder will contract and the other half will bulge until the bubble divides into two.

The film of the soap bubble, which is continually in a state of tension, presses on the enclosed air; by directing the narrow end of the funnel at the flame of a candle you will see that the strength of this very thin film is not so negligible as you might think—the flame wavers quite noticeably (*Fig. 68*).

It is interesting to observe a bubble floating out of a warm room into a cold one. It shrinks noticeably. On the other hand, it expands when brought from a cold room into a warm one. This, naturally, depends on the contraction and expansion of the air inside. If you were to blow a bubble of 1,000 cm³ in a subzero frost of 15°C and then bring it into a room where the temperature is 15°C above zero, it would increase in volume

by roughly 110 cm³ $(1,000 \times 30 \times \frac{1}{273})$.

I must note that a soap bubble is not always as short-lived as is usually thought. When handled with care it can be preserved for some ten days, if not more. The British physicist Dewar, who won fame for his studies of the liquefac-

Fig. 68. The air forced out by the walls of the soap bubble causes the candle-flame to waver

tion of air, preserved soap bubbles in special bottles, well shielded from dust, dryness, and shock, and was able to keep some bubbles for a month and more. The American Lawrence kept soap bubbles under a bell-glass for years on end.

THINNEST OF ALL

Few probably know that the film of a soap bubble is one of the thinnest things you can see with the unaided eye. The customary comparisons we draw upon to express thinness are very thick compared with the film of a soap bubble. A thing "as thin as a hair" or "as thin as cigarette paper" is very thick compared with the walls of a soap bubble, which are 5,000 times thinner than a hair or cigarette paper. A human hair magnified 200 times is about a centimetre thick. If we magnified the cross-section of the film of a soap bubble the same number of times, we still wouldn't be able to see it. We would have to magnify it another 200 times to see it as a slender line. Then a hair—magnified 40,000 times—would be more than two metres thick. *Fig. 69* well illustrates this.

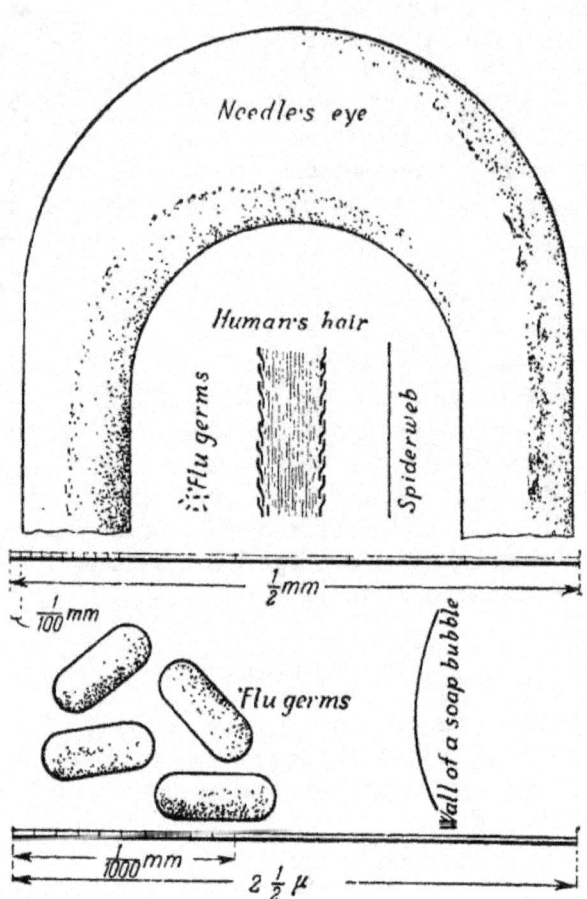

Fig. 69. Top: the eye of a needle, a human hair, germs, and a spiderweb magnified two hundred times. Bottom: germs and the wall of a soap bubble magnified 40,000 times

WITHOUT WETTING A FINGER

Take a large plate and put a coin on it. Then add enough water to cover the coin. Ask your guests to pick up the coin without wetting a finger. It seems impossible, doesn't it?

But it can be solved in a very simple way with the aid of a glass and some paper. Take a piece of paper, light it and, while it is still burning, place it inside the glass. Then quickly put the glass down, bottom up, on the plate. The paper goes out, the glass fills with white wisps of smoke and all the water in the plate flows under it. The coin will naturally remain where it is. A minute or two later, as soon as the coin is dry, you can pick it up without wetting a finger.

What sucked the water under the glass and maintained it there at a certain level? Atmospheric pressure. The burning paper heated the air in the glass, increased its pressure and part of it leaked out. When the paper went out the air cooled again, and its pressure decreased. The pressure of the air outside the glass forced the water in the plate under the glass. Instead of paper you may use matches stuck in a cork as shown in *Fig. 70*.

Fig. 70. How to pick up the coin without wetting a finger

There is current a wrong explanation of this very old experiment (it was first described and properly explained by the physicist Philo of Byzantium who lived somewhere in the 1st century B.C.). Some people say that the water flows under the glass because it is "oxygen that burns out", and that is why the amount of gas in the glass diminishes. This is absolutely wrong. The water flows under the glass only because the air is *heated* and not at all because any oxygen is absorbed by the burning paper. You can check this statement in the following way. Heat up the glass by pouring boiling water into it, thus dispensing with the burning piece of paper. Then, if you take instead of paper a piece of cotton wool soaked in alcohol, which burns longer and heats up the air better, the water will rise up to almost the middle of the glass;

note that oxygen comprises only a fifth of the air in volume. Note, finally, that instead of the allegedly "consumed" oxygen, you have carbon dioxide and water vapour. While the first dissolves in water, vapour remains, replacing part of the oxygen.

HOW WE DRINK

Can this pose a problem? It can. When drinking we bring a glass or a spoonful of liquid up to our lips and suck in the contents. It is this simple thing we are so used to, that we have to explain. Indeed why does the liquid rush into our mouth? What makes it do that? When we drink, our chest expands, thus rarefying the air in our mouth. *The pressure of the outer air forces* the liquid to rush into the place where pressure is less; so does it find itself in the mouth. Liquids in communicating vessels would behave in exactly the same way were we to rarefy the air above one of them. Atmospheric pressure would compel the liquid in this particular vessel to rise. If you enclose the mouth of a bottle with your lips you will fail to suck in the water as the pressure of the air in your mouth and above the water will be the same. So, strictly speaking, we drink not only with our mouths, but also with out lungs, since it is chest expansion that makes the liquid rush into our mouths.

A BETTER FUNNEL

Those who have ever poured liquids into a bottle through a funnel know that from time to time you have to lift the funnel a little because otherwise the liquid will stay in it. This is because the air in the bottle fails to find an outlet and so blocks up the liquid in the funnel. A little of the liquid will drip in so that the air in the bottle is slightly compressed by the liquid's pressure. However, the cramped air will become resilient enough to offset the weight of the liquid in the funnel by its own pressure. By lifting the funnel, we give the compressed air a chance to escape. Then the liquid begins to flow in again. So, to make a better funnel, the narrower part should have ridges outside to prevent the funnel from fitting tightly in the mouth of the bottle.

A TON OF WOOD AND A TON OF IRON

What is heavier—a ton of wood or a ton of iron? Some heedlessly answer that the ton of iron is heavier, thus raising a laugh at their expense. The questioner would probably laugh still louder were he told that the ton of wood is heavier. This seems absolutely incredible, but it is true, strictly speaking.

The point is that Archimedes's principle can be applied not only to liquids but also to gases In the air, every object "loses" in weight as much as the volume of displaced air weighs. Wood and iron also lose a part of their weight, and to get their true weight, you must add the loss. Consequently, the true weight of the wood in our case is one ton plus the weight of the air it displaces.

The true weight of the iron is also one ton plus the weight of the air that the iron displaces. However, a ton of wood occupies a much larger space—about 15 times more—than a ton of iron. Hence, the true weight of a ton of wood is more than that of a ton of iron. Rather should we say that the true weight of the amount of wood which weighs a ton in the air is more than the true weight of iron which also weighs a ton in the air.

Since a ton of iron occupies a volume of $1/8$ m^3 and a ton of wood a volume of about 2 m^3, the difference in the weight of the displaced air should be about 2.5 kg. It is by this amount that a ton of wood is really heavier than a ton of iron.

THE MAN WHO WEIGHED NOTHING

To be as light as a feather—incidentally, in spite of the popular notion, a feather is really hundreds of times heavier than air, and only hovers because due to its rather great "wing-spread" the atmospheric resistance it encounters is much greater than its weight—and even lighter than air, to rid oneself of the fetters of gravity and freely soar into the skies, has been the dream of many a child and even grown-up. But they forget that they can walk around with ease only because they are *heavier than air.*

"We live at the bottom of an ocean of air," Torricelli once said. If we were suddenly to grow a thousand times lighter, lighter than air, we would inevitably float up to the top of this ocean of air. We would rise miles up until we reached regions where the density of the rarefied air would be the same as that of our body. Our dream of hovering in free flight above the hills and vales would be shattered; we would have freed ourselves of gravity but would have been captured by other forces—those of the air currents.

H. G. Wells tells a story in which a very fat man wanted to rid himself of his fatness. The person who tells the story was the possessor of the recipe of a miraculous brew which could rid people of excessive weight. The fat man made the brew according to the recipe and drank it. And this is what happened.

"For a long time the door didn't open.

"I heard the key turn. Then Pyecraft's voice said, 'Come in.'

Fig. 71. "There he was right up close to the cornice"

"I turned the handle and opened the door. Naturally I expected to see Pyecraft.

"Well, you know, he wasn't there!

"I never had such a shock in my life. There was his sitting-room in a state of untidy disorder, plates and dishes among the books and writing things, and several chairs overturned, but Pyecraft—

"'It's all right, o'man; shut the door,' he said, and then I discovered him.

"There he was right up close to the cornice in the corner by the door, as though someone had glued him to the ceiling. His face was anxious and angry. He panted and gesticulated. 'Shut the door,' he said. 'If that woman gets hold of it—'

"I shut the door, and went and stood away from him and stared.

"'If anything gives way and you tumble down,' I said, 'you'll break your neck, Pyecraft.'

"'I wish I could,' he wheezed.

"'A man of your age and weight getting up to kiddish gymnastics—'

"'Don't,' he said, and looked agonised.

"'I'll tell you,' he said, and gesticulated.

"'How the deuce,' said I, 'are you holding on up there?'

"And then abruptly I realised that he was not holding on at all, that he was floating up there—just as a gas-filled bladder might have floated in the same position. He began a struggle to thrust himself away from the ceiling and to clamber down the wall to me. 'It's that prescription,' he panted, as he did so. 'Your great-gran—'

"He took hold of a framed engraving rather carelessly as he spoke and it gave way, and he flew back to the ceiling again, while the picture smashed on the sofa. Bump he went against the ceiling, and I knew then why he was all over white on the more salient curves and angles of his person. He tried again more carefully, coming down by way of the mantel.

"It was really a most extraordinary spectacle, that great, fat, apoplectic-looking man upside down and trying to get from the ceiling to the floor. 'That prescription,' he said. 'Too successful.'

"'How?'

"'Loss of weight—almost complete.'

"And then, of course, I understood.

"'By Jove, Pyecraft,' said I, 'what you wanted was a cure for *fatness*! But you always called it *weight*. You would call it weight.'

"Somehow I was extremely delighted. I quite liked Pyecraft for the time. 'Let me help you!' I said, and took his hand and pulled him down. He kicked about, trying to get foothold somewhere. It was very like holding a flag on a windy day.

"'That table,' he said pointing, 'is solid mahogany and very heavy. If you can put me under that—'

"I did, and there he wallowed about like a captive balloon, while I stood on his hearthrug and talked to him.

" ...'There's one thing pretty evident,' I said, 'that you mustn't do. If you go out of doors you'll go up and up....'

" ...I suggested he should adapt himself to his new conditions. So we came to the really sensible part of the business. I suggested that it would not be difficult for him to learn to walk about on the ceiling with his hands—

"'I can't sleep,' he said.

"But that was no great difficulty. It was quite possible, I pointed out, to make a shake-up under a wire mattress, fasten the under things on with tapes, and have a blanket, sheet, and coverlet to button at the side. He would have to confide in his housekeeper, I said; and after some squabbling he agreed to that. (Afterwards it was quite delightful to see the beautifully matter-of-fact way with which the good lady took all these amazing inversions.) He could have a library ladder in his room, and all his meals could be laid on the top of his bookcase. We also hit on an ingenious device by which he could get to the floor whenever he wanted, which was simply to put the *British Encyclopaedia* (tenth edition) on the top of his open shelves. He just pulled out a couple of volumes and held on, and down he came. And we agreed there must be iron staples along the skirting, so that he could cling to those whenever he wanted to get about the room on the lower level.... (Then, you know, my fatal ingenuity got the better of me.) I was sitting by his fire drinking his whisky, and he was up in his favourite corner by the cornice, tacking a Turkey carpet to the ceiling, when the idea struck me. 'By Jove, Pyecraft!' I said, 'all this is totally unnecessary.'

"And before I could calculate the complete consequences of my notion I blurted it out. 'Lead underclothing,' said I, and the mischief was done.

"Pyecraft received the thing almost in tears. 'To be right ways up again—' he said.

"I gave him the whole secret before I saw where it would take me. 'Buy sheet lead,' I said, 'stamp it into discs. Sew 'em all over your underclothes until you have enough. Have lead-soled boots, carry a bag of solid lead, and the thing is done! Instead of being a prisoner here you may go abroad again, Pyecraft; you may travel—'

"A still happier idea came to me. 'You need never fear a shipwreck. All you need do is just slip off some or all of your clothes, take the necessary amount of luggage in your hand, and float up in the air—' "

At first glance this all seems quite in conformity with the laws of physics. But objections can be made. Firstly, even if Pyecraft had lost his weight, he wouldn't have risen up to the ceiling at all. Recall Archimedes's principle. Pyecraft should have "floated" up to the ceiling only if his clothes and everything in his pockets would have weighed less than the air displaced by his fat body. We can easily reckon the weight of this volume of the air. We weigh almost the same as a similar volume of water—some 60 kg. Air of the usual density is 770 times lighter than water, so the amount we would displace would weigh only 80 gr. However fat Mr. Pyecraft was, he could have scarcely weighed much more than 100 kg; consequently, he must have displaced not more than 130 gr of air. There is no question that Pyecraft's suit, shoes, watch, wallet and all his other belongings weighed more. In that case the fat man should have remained on the floor. He would have felt rather shaky, true, but he certainly would not have "ballooned" up to the ceiling. That would have happened only if he had been stark naked. Dressed, he must have been like a man tied to a bouncing balloon. A small effort, a little jump and he would be up in the air, to smoothly descend again, provided, of course, there was no wind. (See Chapter 4 of my *Mechanics for Entertainment* for more about bouncing balloons.)

"PERPETUAL" CLOCK

You already know a few things about "perpetual motion" machines and of the futility of trying to invent them. Let me now tell you about what I shall call a "gift-power" machine, as it can work indefinitely without human interference, drawing its motive power from the inexhaustible sources of energy in nature. Everybody has most likely seen a barometer, a mercury or aneroid one. In the first one the mercury rises or falls depending on the changes in atmospheric pressure. And it is atmospheric pressure again that causes the arrow to swing in the aneroid barometer.

One 18th-century inventor availed himself of this arrangement to produce a self-winding clock that would never stop. The well-known British mechanic and astronomer James Ferguson saw it in 1774 and this is how he describes it. "I saw this clock," he says, "which is made to go without stopping by the endless rising and falling of the mercury in a curiously arranged barometer. We have no reason to think that the clock would ever stop as the accumulated motive power is enough to make it go for a whole year, even if the barometer were removed. To be frank, I must say that this clock which I examined in detail is the cleverest mechanism I have ever seen, both in design and execution."

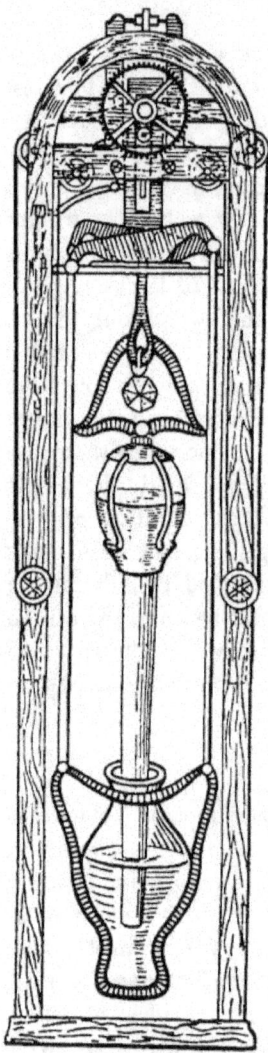

Fig. 72. An 18th-century "gift-power" machine

Unfortunately the clock was stolen and nobody knows what has become of it. Luckily enough, Ferguson made some drawings of it, so it can be reproduced.

Its mechanism consists of a large mercurial barometer, which has about 150 kg of mercury in two glass vessels, one with its mouth in the other, and both suspended in a frame. Both vessels move separately; when atmospheric pressure rises an ingenious system of levers lowers the top vessel and lifts the bottom one. When atmospheric pressure falls, the reverse takes place. This compels a small gear-wheel to turn always in one and the same direction. It doesn't turn only when the atmospheric pressure is steady. However, in these intervals the clockwork is operated by the accumulated potential energy. And though it isn't easy to make the weights rise simultaneously and wind the spring when they drop, the watchmakers of old were ingenious enough.

It even happened that the energy produced by the changes in atmospheric pressure was far more than was needed, causing the weights to rise before they had managed to drop to the bottom. So a special device had to be made to switch off the weights at regular intervals, when they had gone up all the way.

The fundamental difference between such "gift-power" machines and "perpetual motion" machines is obvious. Energy is not produced out of nothing—which was what the inventors of the "perpetual motion" machines sought to achieve. It is supplied from an outside source—in our particular case, the surrounding atmosphere where it is stored up by sunlight. To all practical intents a "gift-power" machine would give the same advantage as could be derived from a "perpetual motion" one—if ever invented—were it not so costly, as it is in most cases.

Later I shall deal with other kinds of "gift-power" machines and shall illustrate why such things are absolutely unprofitable commercially.

CHAPTER SIX

HEAT

WHEN IS THE OKTYABRSKAYA RAILWAY LONGER?

When asked how long the Oktyabrskaya Railway is one person gave this answer: "It's 640 km on the average. But in summer it's about 300 m longer than in winter."

Now this is not so absurd as it may seem. If we meant by the length of a railway the length of its rails, it should indeed be longer in summer than in winter. Don't forget that heat causes steel rails to expand—by more than 100,000th of their length to every one degree Centigrade. On a blazing summer day the temperature of rails might reach 30-40°C and more. Sometimes rails are so hot that they burn the hand. In winter rails may cool down to 25°C below zero and even lower. Supposing that the summer-winter difference in temperature is 55°; by multiplying the railway's total length (640 km) by 0.00001 and again by 55, we get about a third of a kilometre. So in summer the Moscow-Leningrad railway is indeed the third of a kilometre, i. e., roughly 300 m, longer than in winter.

It is, of course, not the length of the railway that changes but merely the sum-total of the lengths of all the rails. This is not one and the same thing, because the rails of a railway track do not directly abut one another. Small spaces are left between their joints for the rails to freely expand when they heat up. (This gap—in the case of 8-metre rails—should be 6 mm at zero. To fully bridge it by expansion the temperature of the rails should rise by 65°C. For certain technical reasons we cannot leave gaps in tramway rails. Usually the rails don't curve, because they are sunk in the ground, temperature fluctuation is not so great and the method used to spike the rails prevents them from curving. However, on a very hot day tram rails do curve, as *Fig. 73*, the reproduction of an actual photograph, well illustrates. Sometimes the same thing hap-

Fig. 73. Tram rails bend on very hot days

pens to the rails of a railway track. On downgrades the train pulls at the rails—sometimes even together with the sleepers. As a result, the gaps often disappear on such sections and the rails directly abut one another.) The calculation we have made shows that the total length of all the rails increases at the expense of the total length of these gaps; on a hot summer day the total length in our particular case is 300 metres more than in a winter frost. So to sum up: the rails of the Oktyabrskaya Railway are indeed 300 m longer in summer than in winter.

UNPUNISHED THEFT

On the Moscow-Leningrad line several hundred metres of costly telephone and telegraph wire vanish without trace every winter. Nobody is ever worried; all know who the culprit is. I suppose you too have guessed by now. The thief, of course, is the frost. What is true for rails is true for wire too. The only difference is that copper telephone wires expand 1.5 times more than steel, when heated. And since we have no gaps here we can really say, without any reservations whatsoever, that *in winter the Moscow-Leningrad telephone line is indeed 500 m shorter than in summer.* Every winter the frost steals nearly half a kilometre of wire and gets away with it! But it doesn't disrupt telephone or telegraph communications. All that is stolen is dutifully refunded when warmer days set in.

But when bridges, not wires, contract due to frosts the consequences are pretty bad. Newspapers had this to report in December 1927: "The unusual frosts France has been having lately have seriously damaged the bridge across the Seine in the heart of Paris. Due to frosts the bridge's steel framework contracted, causing the road blocks to fly out. The bridge has been temporarily closed to traffic."

HOW HIGH IS THE EIFFEL TOWER?

If I were to ask you now how high the Eiffel Tower is, before saying "300 metres", you would probably want to know in what weather—cold or warm? After all, the height of such an enormous steel structure could not be the same at all temperatures. We know that a steel rod 300 m long expands by 3 mm when heated by 1° C. The height of the Eiffel Tower should increase by roughly the same amount when the temperature rises by 1°. In warm sunny weather the steel framework of the tower might warm up in Paris to 40°C above zero, whereas on a cold rainy day its temperature might fall to 10°C and in winter down to zero and even to as much as 10° below (heavy frosts are rare in Paris). The temperature fluctuation is as much as 40° and more. This means that the height of the Eiffel Tower may be 3 × 40 = 120 mm = 12 cm more or less.

Direct measurement has disclosed that the Eiffel Tower is still more sensitive to temperature fluctuations than the air itself. It warms up and cools quicker and reacts sooner to the sun's sudden appearance on a cloudy day. The changes in the height of the Eiffel Tower were detected by using a wire made of a special nickel steel on whose length temperature fluctuations have practically no effect. This wonderful alloy is called invar from the word invariable.

So, on a hot day the Eiffel Tower is taller than on a cold day by a bit equal to 12 cm and made of iron, which, incidentally, doesn't cost a sou.

FROM TEA GLASS TO WATER GAUGE

Before pouring tea into a glass, the experienced housewife puts in a tea spoon, especially a silver one, to prevent the glass from cracking. Practice has suggested the proper solution.

But what is its basic principle? Why does hot water crack a tea glass?

Because of the uneven expansion of the glass. When you pour hot water into a glass, not all its walls warm up at once. At first the inner layer warms up, the outer one remaining cold. The heated inner layer expands at once. Meanwhile, since the outer one does not expand, it feels a strong pressure from inside. It snaps and the glass breaks.

Don't think you can safeguard yourself against this by using thick-walled glasses. They, on the contrary, are liable to crack sooner than thin-walled ones. This is because a thin wall heats up faster and its temperature and expansion even out sooner. A thick-walled glass, on the other hand, warms up slowly.

One thing you mustn't forget when buying thin-walled glassware—make sure that the bottom of the glass is thin too, because it is the bottom that chiefly heats up. A thick-bottomed glass will crack, however thin its walls. So do glasses and china cups with thick-rimmed bottoms.

The thinner-walled a glass vessel is, the safer it is for heating. Chemists use very thin-walled vessels in which they boil water right over the burner.

The ideal vessel is one that wouldn't expand at all when heated. Quartz almost has this property: it expands 15-20 times less than glass. A thick-walled vessel of transparent quartz will never crack when heated, even if immersed red-hot in a bath of ice (vessels of quartz are good for laboratory work because it melts only at 1,700°C). This is also partially because quartz conducts heat much better than glass.

Tea glasses crack not only when warmed up quickly but also when cooled quickly. Now it is uneven contraction that is to blame. As it cools, the outer layer contracts and exerts a strong pressure on the inner layer, which has not cooled and contracted yet. A prudent housewife should not put a jar of hot jam out in the cold or into cold water.

But back to the tea spoon. How does it protect the glass from cracking? The difference in the expansion of the inner and outer layers is great only when very hot water is poured into the glass at once. Warm water, however, doesn't make glasses crack. What happens when you put a tea spoon in? As it pours in, the hot water loses part of its heat to the metal spoon, which is, contrary to glass, a good conductor of heat. Its temperature drops and it becomes almost harmless, because now it is only warm. Meanwhile the glass has warmed up and more hot water won't crack it.

In a nutshell, a metal tea spoon, especially a heavy one, offsets the uneven heating of the glass and prevents it from cracking.

But why is a silver spoon still better? Because silver is a very good conductor of heat. It can take away the heat from the water sooner than a copper spoon. A silver spoon in a glass of hot tea burns the fingers. Since a copper spoon doesn't do that, you can easily tell the material the spoon is made of.

The uneven expansion of glass walls is a menace not only to tea glasses but also to very important elements of boilers—the water gauges which give the height of the water in the boiler. As the hot steam and water heat them up, their inner layers—they are tubes of glass—expand more than their outer layers. Add to this the great pressure exerted in the tubes by the steam and water, and you will realise why they may so easily burst. To prevent this, they are sometimes made of two layers of different kinds of glass, the inner one having a smaller expansion factor than the outer one.

THE BOOT IN THE BATHHOUSE

"Why in winter is the day short and the night long, and in summer the other way round? The winter day is short because like all other visible and invisible things it contracts due to cold; meanwhile the night expands—it is warmed up when lights and lamps are lit." How comically silly this "explanation", afforded by Chekhov's retired Don

Cossack sergeant, is. However, people who ridicule such "learned" reasoning sometimes father theories which are just as stupid. Have you ever heard the story of the boot which won't go on in the bathhouse because "the heated foot has grown larger"? A classical instance, but with a totally wrong explanation.

In the first place one's temperature hardly rises at all when one is in a bathhouse—never by more than one degree Centigrade. Only a Turkish bath will make it go up two degrees. Our body successfully resists the surrounding heat, maintaining its temperature at a definite level. Furthermore, this "rise" in our body temperature increases the volume of our body by such a negligible fraction that one doesn't notice it when drawing on a boot. The expansion factor of our bones and flesh is never more than a few ten-thousandths. Consequently, the sole and the instep could bulge only by a hundredth of a centimetre—no more. Boots and shoes are never sewn with such accuracy. After all, a hundredth of a centimetre is but the thickness of a hair!

Still it remains a fact that it is hard to draw a boot on after a hot bath. However, this is not because our foot expands due to heat but because the blood rushes to the foot, the skin swells, is damp, and grows tender—in a word, because of things that have nothing at all in common with expansion due to heat.

HOW TO WORK MIRACLES

Hero of Alexandria, the ancient Greek mathematician who invented the fountain that bears his name, has left the description of two artful methods which enabled Egyptian priests to take in worshippers by their "miracles".

Fig. 74 shows one such device consisting of a hollow metal altar which stood in front of the temple doors, and of the mechanism, hidden beneath the flagstones, that caused the temple doors to open. When incense was burned, the heated air inside the hollow altar exerted a greater pressure on the water in the vessel hidden below the floor, thus causing it to flow through a pipe into a pail which lowered and set in motion the door-opening mechanism (*Fig. 75*). The worshippers saw, of course, what they thought to be a "miracle"—the temple doors swung

Fig. 74. Egyptian temple "miracle" explained. The doors open when incense is burned on the altar

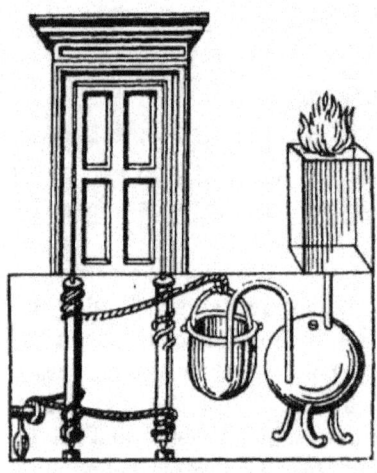

Fig. 75. Diagram showing how the temple doors swing open. (Compare with *Fig. 74.*)

Fig. 76. Another fake miracle of the ancient priests. How incense "everlastingly" drips into the sacrificial flame

open of their own accord as soon as incense and prayers were offered by
the priests. They, naturally, knew nothing of the hidden mechanism.

Another fake "miracle" which the priests staged is shown in *Fig. 76*.
As soon as incense is burned the expanding air forces more of it to flow
out of the cistern below the floor into pipes concealed inside the figures
of the priests. The worshippers beheld the "miracle" of an undying
flame. However, when the priest in charge considered the offerings too
scanty, he unnoticeably removed the stopper in the lid of the cistern.
This stopped the flow of incense, because now the superfluous air could
find a free outlet.

SELF-WINDING CLOCK

At the close of the previous chapter I described a self-winding clock;
its working principle was based on the changes in atmospheric pressure.
Now I shall tell you about similar self-winding clocks, the principle
of which is based on heat expansion. *Fig. 77* depicts the mechanism of
one of them. The central element
consists of rods Z_1 and Z_2 which
are made of a special alloy with a
considerable coefficient of expansion. Upon *expansion* rod Z_1 engages
the teeth of wheel X, turning it.
Upon *contraction*, on the other
hand, rod Z_2 engages the teeth of
the wheel Y, turning it in the same
direction. Both wheels are set on
shaft W_1 which also revolves a large
wheel with scoops on it. These
scoops lift the mercury from the
lower inclined tank R_1 to another contrarily-inclined tray R_2

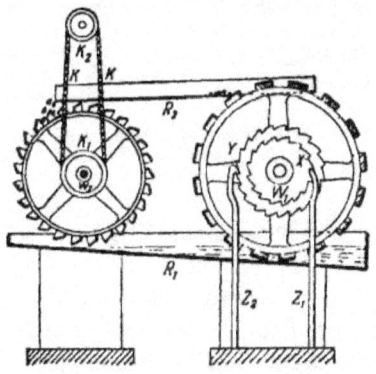

Fig. 77. Diagram of a self-winding
clock

down which it flows towards the left-hand wheel also with scoops.
As these scoops fill, the wheel turns, setting in motion chain KK,
looped around wheel K_1, which is set on the same shaft W_2 as the
big wheel, and around wheel K_2, which winds up the clock. Meanwhile
the scoops of the left-hand wheel spill out the mercury into the inclined

tank R_1, down which it flows to reach the right-hand wheel, and the cycle begins all over again.

This clock, apparently, would go on ticking, while rods Z_1 and Z_2 expand and contract. All we need to wind the clock is an alternate rise and fall in air temperature, which is something that takes place without our interference. Could we call this clock a "perpetual motion" machine

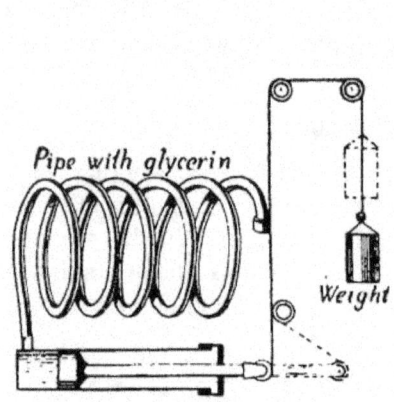

Pipe with glycerin

Weight

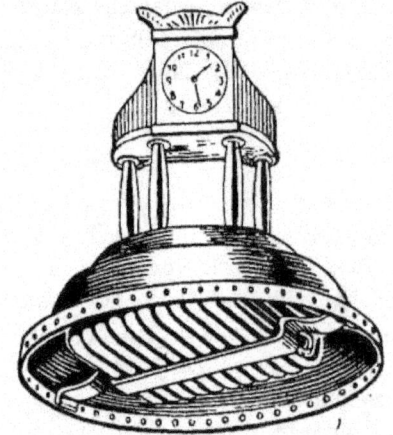

Fig. 78. Diagram of another self-winding clock

Fig. 79. Self-winding clock. The pipe with the glycerin is hidden in the base of the clock

then? Of course not. The clock will tick indefinitely until its mechanism wears out, but what makes it go is the heat of the surrounding air. The clock stores up the work of heat expansion and expends it portion after portion to turn its hands. This is really a "gift-power" machine since it does not require care or outlay. But it doesn't create energy out of nothing; its primary source is the heat of the sun, which warms up the earth.

Another specimen of a self-winding clock with a similar arrangement is given in *Figs. 78* and *79*. Its basic element is glycerin, which expands when the temperature of the air rises and causes a small weight to rise. The lowering of this weight makes the clock go. Since glycerin solidifies only at 30°C below zero and boils at 290°C above, this mechanism is quite suitable for town clocks. A 2° temperature fluctuation is already enough to keep it going. One such clock was tested for a whole year, and proved to be quite satisfactory.

Can any advantage be derived by designing other bigger machines of this kind? At first glance, such a "gift-power" machine might seem very economical. Let us see, though, whether this is really so. To wind up an ordinary clock to run for 24 hours one requires only 1/7 kgm of energy. This is merely $\frac{1}{600,000}$ of a kilogramme-metre per second. Considering that one horsepower is equivalent to 75 kgm/sec, the power of one clock mechanism is equivalent to only $\frac{1}{45,000,000}$ of a horsepower. Consequently, if the rods in the first clock mentioned or the contraption of the second were to cost one kopek, the investment made to produce one h.p. would be 45,000,000 kopeks, or 450,000 rubles. I think half a million rubles for one horsepower is a bit too much for a "gift-power" machine.

INSTRUCTIVE CIGARETTE

Fig. 80 shows a straw-tipped cigarette on top of a match box. Smoke is curling out of both ends. However, at one end it *curls up*, and at the other *down*. Why? After all, isn't the smoke coming out of the two ends the same? It is, of course, but above the smouldering end there is an ascending current of warm air which carries the particles of smoke up. Meanwhile the air carrying the smoke through the straw tip cools off and no longer rises upward; since the particles of smoke are heavier than air, they float down.

ICE THAT DOES N'T MELT IN BOILING WATER

Take a test tube, fill it with water, and put a lump of ice in. To keep the ice down at the bottom—since it is lighter than water, it floats—press it down by some small weight, seeing to it that the water can get at the lump of ice. Now heat the test tube on a spirit lamp so that the flame licks

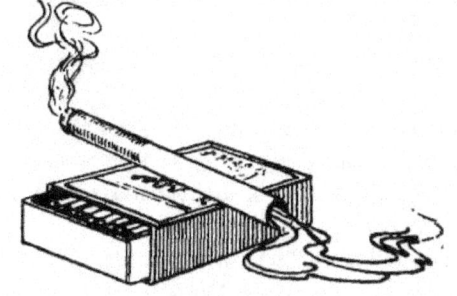

Fig. 80. Why does the smoke curl up from one end, and down from the other?

Fig. 81. The water at the top boils, but the ice at the bottom doesn't melt

only at the tube's upper part as shown in *Fig. 81.* The water will soon boil and send out steam. Oddly enough, the ice at the bottom of the tube doesn't melt. A minor miracle, one would think—ice that doesn't melt in boiling water!

The trick is that at the bottom of the tube the water doesn't boil at all; it remains *cold*. Actually we have not "ice in boiling water" but "ice beneath boiling water". As it expands due to heat, the water becomes lighter; it does not descend to the bottom and stays in the upper part of the tube. There is warm water and a mixture of warm and cold layers of water only in the tube's upper part. Heat can be transferred down only by a conductor, but water is a very poor conductor of heat.

ON TOP OR BENEATH?

When we want to heat water, we put the vessel that contains it right above the flame and not to the side of it. This is the right thing to do since the heated air which grows lighter is forced out from beneath the vessel *upwards* and thus envelops the vessel. So by placing the object we want to heat up right above the flame we use the source of heat in the most advantageous way.

But what should we do *to cool* something with ice? Many put the thing they want to cool—a jug of milk, for example—on top of the ice. This is the wrong thing to do; as the air above the ice cools it *descends*, its place being taken by the warmer surrounding air. So if you want to cool a drink or a dish, *don't put it on top* of the ice but rather the *ice on top of it.*

Let me make the point clearer. When we put a jar of water on top of ice, it is only the bottom layer that cools. The rest of the water is surrounded by uncooled air. But if we put the ice *on* the lid, the water will cool much faster. The cooled upper layers will descend, their place being taken by the warm layers rising from the bottom; the process goes on until all the water has cooled (note that pure water will cool not to zero but only to 4 °C above—the temperature at which it possesses the greatest density. After all we never really cool drinks down to zero). Meanwhile the cooled air around the ice will also descend and envelop the vessel.

DRAUGHT FROM CLOSED WINDOW

We often feel a draught coming from a window that is closed tight and hasn't a single crack in it. Though it seems odd there is nothing at all surprising in it.

The air inside a room is practically never in a state of rest. An invisible current circulates as the air warms or cools. As the air warms it rarefies and grows lighter. As it cools it becomes denser and heavier.

The cold heavy air near the windows and outer wall descends to the floor, forcing the warm light air to rise to the ceiling. A toy balloon reveals this circulation at once. Tie a small weight to it, light enough to keep it suspended in mid-air. Release the balloon near the stove or radiator. You will see it travel around the room, being carried by the invisible current from the fireplace or radiator up to the ceiling and towards the window, and from there down to the floor and back to the fireplace. Here it again sets out on the same journey. That is why we feel the draught, especially around the feet, coming from the window though it is closed tight in winter.

MYSTERIOUS TWIRL

Take some thin cigarette paper and cut out a piece in the form of a rectangle. Fold it down the middle and then straighten it again. The fold will tell you where the centre of gravity is. Now stick a needle upright into the table and place the piece on the other end so that it is set on its centre of gravity and, hence, balanced. So far there is nothing

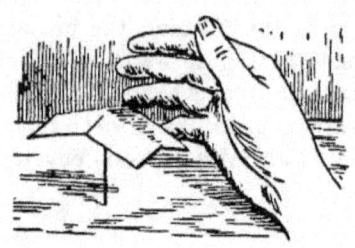

Fig. 82. Why does this piece of paper spin?

mysterious about it. Bring up your hand as is shown on *Fig. 82*. Do this gently though, otherwise the piece of paper will be blown off by the rush of air. The paper will start to spin. At first it gyrates slowly but then it picks up speed. Take your hand away and gyration stops. Bring your hand up again and gyration resumes.

This mysterious gyration once—in the 1870's—caused many to believe that we, or rather our bodies, were endowed with some supernatural properties. Mystics thought this confirmed their wild theories about the strange fluids the human body was supposed to possess. Actually, there is nothing unnatural in it; as a matter of fact, everything is as simple as pie. When you bring your hand up, the air near it, which is warmed by its proximity, rises and, pressing against the piece of paper, causes it to spin. It revolves because it is slightly folded, thus acting the same role as a curled piece of paper suspended above a lamp.

A closer look will show you that the piece of paper always gyrates in one and the same direction—from the wrist towards the finger-tips. This is because the finger-tips are always colder than the palm of the hand; consequently, the palm gives rise to a stronger ascending air current than the finger-tips. Incidentally, when one is feverish, or happens to be running a high temperature, the paper gyrates much faster. You might be interested to learn that this twirling, which once mystified so many, was the subject of a communication made to the Moscow Medical Society in 1876 (*The Gyration of Light Bodies Caused by the Heat of the Hand*, by N. P. Nechayev).

DOES A WINTER COAT WARM YOU?

If I told you that your fur coat *does not warm* you at all, you would probably think I was pulling your leg. But suppose I prove it? Stage the following experiment. Take the reading of an ordinary thermometer.

Then wrap it in your fur coat and let it be for some hours. Then read the thermometer again. It will be exactly the same as before. Has that convinced you that your fur coat doesn't warm you? Perhaps, it *cools* you then? Take two bags of ice and wrap one in your fur coat, leaving the other in a dish. When this second bag of ice melts, unwrap the coat. The ice in the first bag has hardly melted at all. As you see, the coat has not warmed it in the least; on the contrary, it seems even to have cooled it, since the ice took longer to melt!

So, does a winter coat warm you? No, if by warming we mean the *communication of heat*. A lamp does. So does a stove. And so does our body. They are all sources of heat. Your fur coat is not a source of heat; *it doesn't have any warmth of its own to give. It merely prevents our body from shedding its own warmth.* That is why a warm-blooded animal —whose body is actually a source of heat—feels much warmer in a coat of fur than without one. However, since the thermometer we took for our experiment is not a source of heat its reading naturally could not change simply because we wrapped it in the fur coat. The ice in the coat also took longer to melt because the coat is a rather poor conductor of heat and blocks any intake of surrounding warmth.

The snow on the ground is also like a fur coat; it is a poor conductor of heat—like all powdery bodies—and thus prevents the ground beneath from shedding its heat. The temperature of the ground beneath a protective layer of snow is often some 10°C higher than at a bare spot.

So the answer to the question "Does a winter coat warm you?" is: it merely helps us to warm ourselves; rather we ourselves warm the coat instead.

THE SEASON UNDERFOOT

It is summer on the ground and above it. What season of the year is it three metres down? You think it's summer? You're wrong! It's not at all the same season as one might think. The point is that the ground is a very poor conductor of heat. In Leningrad water mains don't burst even in the grimmest of frosts, because they are two metres deep. Above-surface temperature fluctuations reach the different subsoil strata with great delay. Direct measurements conducted in the town of Slutsk, Leningrad Region, showed that at three metres down the warmest time of the year comes 76 days late, while the coldest period is 108 days late. If the hottest day above the ground is July 25th, at three metres down the hottest day will come only on October 9th. On the other hand, if the coldest day is January 15th, at the depth given the coldest day will come only in May. At greater depths the delay is still greater.

The further down we go, the weaker the temperature fluctuations become, to fade to an everlasting constant at a certain depth; here you have one and the same temperature round the year for centuries on end. This temperature is the mean annual temperature of the place in question. In the cellars of the Paris observatory, 28 metres below the ground, there is a thermometer which Lavoisier stored away there more than 150 years ago. The mercury has not budged a hair since, giving all the time one and the same temperature of 11.7 °C above zero.

To sum up: underfoot we never have the season of the year we have above the ground. When it is winter for us it is still autumn three metres down—of course not the autumn we had, as the fall in temperature is not so pronounced. On the other hand, when it is summer for us, deep down we still have faint repercussions of winter frosts. One must always bear this important point in mind whenever one is dealing with the conditions of life underground—for plant tubers and roots, and for cockchafer grub, for instance. It should not be surprising, for instance, that in tree roots the cells multiply in winter and that the tissue called the cambium ceases to function for practically the whole of summer, in contrast to the tissue of the above-ground tree-trunk.

PAPER POT

Look at *Fig. 83*. An egg is boiling in water in a paper cup. Won't the paper burn through and the water spill out and extinguish the flame? Try to do it yourself. boiling the egg in some stiff parchment paper attached fast to a piece of wire (or better make the paper box shown in *Fig. 84*). Nothing happens to the paper! The reason is that one can warm water only up to boiling point—100 °C. The water—it has a great capacity for absorbing heat—absorbs the paper's extra heat and prevents it from warming to much more than 100 °C, that is, to

a point where it could burst into flame. The paper won't burn even if licked by the flame.

It is the same property of water that prevents a kettle from

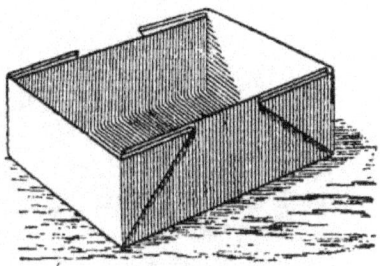

Fig. 83. An egg boiling in a paper pot   Fig. 84. A paper box for boiling water

going to pieces—which is what would happen were we absent-minded enough to put the kettle on to boil without any water in it. For the same reason you must not put soldered pots on the fire unless they have water in them. The water used to cool the old Maxim machine guns saved the barrel from melting.

By using a little box made from a playing card, you can melt a lead pellet. To do this, put the lead in the box right above the flame. Since lead is a good conductor of heat it rapidly absorbs the heat of the box, preventing the box from heating up to way above its melting point—335 °C—which is too little yet for the box to break into flame.

Fig. 85 gives another simple experiment. Take a thick nail or an iron—or better copper—rod and *tightly* curl screw-wise a narrow strip

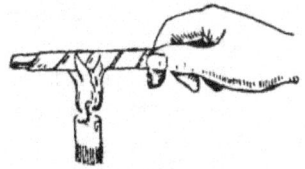

Fig. 85. Paper that doesn't burn

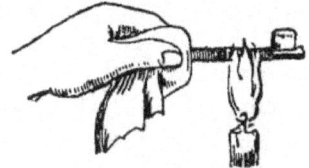

Fig. 86. The thread that doesn't burn

of paper around it. Then apply a flame. The flame will lick at the paper and even smoke it; but it'll start burning only when the rod grows red-hot. Again the metal's good heat conductivity is the reason. A glass stick, for instance, wouldn't do at all for this experiment. *Fig. 86* shows a similar experiment in which we have a "non-inflammable" piece of thread wound *tightly* round a key.

WHY IS ICE SLIPPERY?

One slips on a smoothly polished floor much more easily than on one that isn't polished. Now, shouldn't *smooth* ice be much more slippery than bumpy ice? However, contrary to expectation, a sled goes much more easily over bumpy ice than over smooth ice—which you may have noticed yourself if you have ever happened to pull a sled. How come that bumpy ice is more slippery than glossy ice? Ice is slippery not because it's smooth but because its melting point drops when pressure is increased.

Let's see what happens when we sled or skate. On skates we bring the whole weight of our body to bear down on a very small area, of but a few square millimetres. Recall Chapter 2 of this book. You will realise that a person on skates exerts a considerable pressure on the ice. Under strong pressure ice melts at a lower temperature. For instance, if the temperature of the ice is 5 °C below zero and the skater's pressure has lowered the melting point of the ice beneath his skates by 6 or 7°, this ice will melt. This gives rise to a thin layer of water between the blades and the ice. No wonder the skater slides, or rather slips, along. And as soon as he moves further, the same thing repeats itself. The skater continually slides over a thin layer of water. It is only ice that has this property. One Soviet physicist even called it "nature's sole slippery body". All other bodies are smooth but not slippery.

Back now to our first point. Why is bumpy ice more slippery than smooth ice? We already know that one and the same weight exerts a stronger pressure when it rests on a smaller area. When does a man exert more pressure? On smooth ice? Or on bumpy ice? It is quite obvious that he exerts more pressure on bumpy ice because in this case he is supported only by a few bumps in the ice. The greater the pressure exerted, the more readily does ice melt and, consequently, the more slippery does ice become—provided the sled runners are wide enough (this will not apply to the thin skate blades as the energy of motion is expended to slice off the bumps).

This pressure-induced lowering of the melting point of ice explains many other things that we see around us. This is why separate lumps of ice freeze into one when strongly pressed together. Boys throwing snowballs unconsciously avail themselves of this property; the separate snowflakes stick together because the pressure exerted to form the snowball lowers their melting point. To make a snowman we again apply this principle. (I suppose I needn't explain, though, why in strong frosts we are unable to mould good snowballs and snowmen.) Under the pressure of the many feet walking along the pavement snow gradually turns into one solid icy mass.

It has been theoretically calculated that to lower the melting point of ice by one degree Centigrade we must exert the rather considerable pressure of 130 kg/cm². Here one must bear in mind that in the process of melting both ice and water are subjected to one and the same pressure. In the instances described it was only the ice that was subjected to strong pressure; the water the ice melted into is subjected to atmospheric pressure; consequently, in this case the effect pressure has on the melting point of ice is much greater.

THE ICICLES PROBLEM

Have you ever stopped to wonder how the icicles we see drooping from eaves form? And when do they form? During a thaw or during a frost? And if during a thaw, then how does water freeze at an above-zero temperature? On the other hand, if during a frost, then where, in general, does the water that freezes come from?

As you see, the problem is not so simple as you may have thought. To produce icicles you need *two temperatures* simultaneously—one above zero for melting and the other below zero for freezing. That is really what happens. The snow on slanting rooftops melts because it is warmed by the sun to an *above-zero* temperature. Meanwhile the drops of water dripping off the eaves freeze, because here we have a *sub-zero*

Fig. 87. The sun heats the slanted roof more than the ground

temperature. (We don't mean the icicles that form because of the warmth exuded by the heated room under the roof.)

Try to imagine the following picture. It's a clear and sunny day. The temperature is just one or two degrees Centigrade below zero. Everything is bathed in sunlight. The sun's slanting rays are not strong enough to melt the snow on the ground. But since they strike the *inclined* rooftop facing the sun *at an angle closer to a right angle*, they warm up the roof and. melt the snow on it. Sunshine gives more light and warmth the wider the angle between the line of the rays and the plane on which they are incident. It acts in direct ratio to the *sine* of this angle. As for the case in *Fig. 87*, the snow on the rooftop gets 2.5 times more warmth than the snow on the ground, because the sine of 60° is 2.5 times more than the sine of 20°. The melting snow drips off the eaves. But since the temperature beneath the eaves is a *sub-zero* one the drops of water—cooled furthermore by evaporation—freeze. Another drop drips onto the frozen one and also freezes. Then comes a third, a fourth and so on, gradually producing a tiny pendant of ice. A couple of days later, or maybe a week later, we have the same kind of weather again. The pendant grows, producing a larger and larger icicle—in much the same way as lime stalactites form in underground caverns. That is how icicles form on the eaves of sheds and other unheated premises.

The changing angle of incidence of the sun's rays produces far grander phenomena. The different climatic zones and seasons are largely due to that—but not wholly; another major factor is the varying day-length, or the time during which the sun warms the earth, which, like the seasons, is due to one and the same astronomical cause, the inclination of the earth's axis of rotation to the plane of the ecliptic. In winter the sun is practically as far away from us as in summer; it is just as far away from the poles as it is from the equator—the difference is so insignificant that it can be totally ignored. However, at the equator the angle of incidence of the sun's rays is wider than at the poles; in summer again, the angle of incidence is wider than in winter. This phenomenon gives rise to a pronounced variation in temperatures, and consequently in nature in general.

CHAPTER SEVEN

LIGHT

TRAPPED SHADOWS

Our forefathers did find some use for their shadows even though they weren't able to catch them. This was the making of silhouettes, or shadow images. Today we go to the photographer's if we want our pictures or the pictures of friends and relatives taken. But in the 18th century there were no photographers. Portrait-painters asked a stiff price for their work and only the rich could afford it. That is why *silhouettes* were so widespread; in some measure they did for our present snapshots.

Silhouettes are actually trapped shadows. They were obtained mechanically and in this we can draw a certain parallel between them and their opposites—photographs; while photographers draw on *light* ("photos" is Greek for light) to make pictures, our ancestors used *shadows* for the same purpose.

Fig. 88 shows you how silhouettes were made. The sitter turned his head to cast a characteristic profile and this profile was traced with a pencil. Then the inside of the outline was blacked, cut out, and glued onto a white ground. This was the silhouette. Whenever necessary, the silhouette was reduced by means of a special device called the pantograph (*Fig. 89*).

Don't think that this simple black outline could not give a notion of the characteristic features and profile of its prototype. A good silhouette is sometimes amazingly like the original.

This property intrigued some artists, who began to paint in this manner, thus starting a whole school. The very origin of the word is of interest. It derives from Etienne de Silhouette, an 18th-century

Fig. 88. An old way of making shadow portraits

Fig. 89. How to reduce a silhouette

Fig. 90. A silhouette of Schiller (1790)

French Minister of Finance, who urged his extravagant compatriots to show thrift and reproached the French aristocracy for wasting money on pictures and portraits. The cheapness of shadow likeness thus suggested the name—portraits "à la Silhouette".

THE CHICK IN THE EGG

The properties shadows possess will enable you to stage an amusing parlour trick. Take a piece of greased paper and make a screen by sticking it on top of a square hole cut in a piece of cardboard. Put two unshaded table lamps behind this screen and seat your friends in front of it. Switch on the left lamp. Place an oval piece of cardboard mounted on a piece of wire between the lit lamp and the screen. Your friends will naturally see the outline of an egg. The second lamp is still not on. Now tell your friends that you have an X-ray machine that will detect the chick inside the egg. Hey, presto! and your friends see the egg's shadow pale and the rather distinct outline of a chick appear in the middle (*Fig. 91*).

It is really all very simple. Just switch on the right lamp which has a cardboard chick between it and the screen. Part of the oval shadow upon which the chick's shadow is superimposed is illumined by the right lamp. That is why its fringes are lighter. Since your friends don't see your manipulations, those ignorant of physics and anatomy may really think that you have X-rayed the egg.

Fig. 91. A fake X-ray

PHOTOGRAPHIC CARICATURES

Many of you might not know that you can make a camera in which an ordinary small round hole will take the place of the lens. True, you get a fainter image in this case. An interesting modification of this "lensless" camera is the "slit" camera which has two criss-crossing slits instead of the round aperture. This camera has in its front part two small slats, one having a vertical slit and the other a horizontal slit. When the two slats are superimposed the image obtained is the same as produced by the aperture camera. In other words, the likeness is not distorted. But when the slats are moved apart—they are specially arranged so that this can be done—the image produced becomes distorted (*Figs. 92* and *93*), resembling a caricature rather than a photograph.

Fig. 92. A caricature obtained by means of a "slit" camera. The image is distended horizontally

Fig. 93. A similar caricature distended vertically

Why does this happen? Let us take the case when the slat with the horizontal slit is placed in front of that with the vertical slit (*Fig. 94*). The rays coming from the vertical line of figure D (a cross) pass through the first slit C as through any ordinary aperture; meanwhile slit B does not alter their course at all. Consequently, on the ground-glass screen A you get an image of the vertical line on a scale corresponding to the distance between A and C. However, this disposition of the slats produces an entirely different image of D's horizontal line. The rays pass through the horizontal slit without hindrance and don't cross until they reach the vertical slit B, which they pass as any round aperture to produce on screen *A* an image on a scale corresponding to the distance between *A* and *B*.

In short, the vertical lines are taken care of by slit *C* only, and the horizontal lines, on the contrary, by slit *B* only. Since slit *C* is further away from the screen all vertical dimensions are reproduced on glass *A* on a scale larger than that of the horizontal dimensions. In other

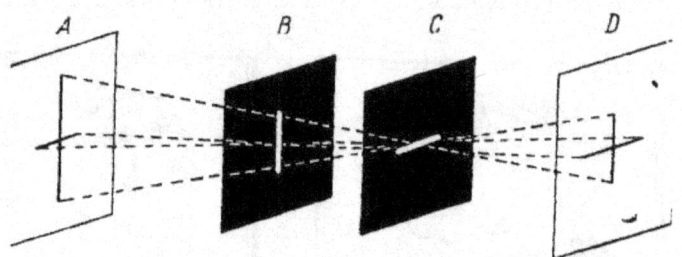

Fig. 94. Why the "slit" camera produces distorted images

words the image is distended vertically. A redisposition of the slats will produce a horizontally distended likeness (compare *Figs. 92* and *93*). A *slantwise* disposition will distort the likeness in still another way.

This camera can be employed not only to get caricatures. It can also serve a more serious purpose, as, for instance, to vary architectural embellishments, carpet and wallpaper patterns, and in general any ornamental motif that may be distended or condensed at will in a definite direction.

THE SUNRISE PROBLEM

Suppose you get up exactly at 5 o'clock early in the morning to watch the sunrise. Since light does not propagate instantaneously some time must pass before the light reaches your eye from its source. So my question is: At what time would you see the sunrise were light able to propagate instantaneously?

Since it takes eight minutes for the light to travel from the sun to us here on Earth, one might think that if light propagated *instantaneously* one would see the sun rise eight minutes earlier—at 4:52 a.m. You're in for a surprise if you think so; that answer is absolutely wrong. The sun "rises" when the Earth turns to face the *space that is already lit.* Therefore even if light propagated instantaneously we would still see the sunrise only at 5 a.m.

If we take what is called "atmospheric refraction" into consideration we get a still more startling result. Refraction curves the path of light, thus enabling us to see the sun "rise" *before* it really rises above the horizon. But if light propagated instantaneously, there would be no refraction as this is due to the different velocities with which light travels in different media. And as there would be no refraction, we would see the sun rise a bit *later*—from two minutes to as much as several days and even more (in polar latitudes), as this would depend on the latitude, air temperature, and certain other factors. So, were light to propagate instantaneously we would see the sunrise later than we do now. A most curious paradox! (See *Do You Know Your Physics?* for further detail.)

It would be quite different, of course, if you were observing the appearance of a solar protuberance in a telescope. Then—that is, if light propagated instantaneously—you would see it eight minutes earlier.

CHAPTER EIGHT

REFLECTION AND REFRACTION

SEEING THROUGH WALLS

In the 1890's one could buy a curious contraption pompously called an "X-ray apparatus". I remember how puzzled I was when I, a schoolboy at the time, saw this ingenious device for the first time. It enabled me to see light through opaque objects—not only thick paper but even a knife blade, which is impenetrable to real X-rays. *Fig. 95*, which shows the prototype of the contraption I just mentioned, "lets the cat out of the bag". It has four small mirrors, each slanted at the angle of 45°, to reflect and rereflect the rays coming from the object and thus lead them around the opaque obstacle.

Fig. 95. A sham X-ray apparatus

The military extensively employ a similar device—the periscope (*Fig. 96*)—enabling them to follow the enemy's movements without exposing themselves to the hazard of enemy fire. The further away the

Fig. 96. The periscope

Fig. 97. Diagram of a submarine periscope

object is from the periscope, the smaller the observer's field of vision is. A special arrangement of optical lenses is used to enlarge the field of vision. But since the lenses absorb part of the light that enters the periscope, the image obtained is blurred. This limits the height of a periscope, with some twenty metres being already close to the "ceiling". Taller periscopes give a very small field of vision and a blurred image, especially in cloudy weather.

Submarine commanders also use periscopes to watch the ships they attack. Though a far more complicated affair than the army periscope, this periscope, which juts out of the water when the submarine submerges, is the same in principle, having a similar arrangement of mirrors (or prisms). (*Fig. 97.*)

THE SPEAKING HEAD

This frequent side-show "marvel" dumbfounds the uninitiated. It does, indeed, astound one to see on a plate a live, seemingly severed human head, which rolls its eyes, speaks, and eats. And though you can't walk right up to the table on which it lies, you "quite perfectly" see that there is nothing underneath. If you ever see this side show, make a paper ball and throw it under the table. Strangely enough, it bounces back. The mystery is no longer a mystery—it has bounced off a mirror. Even if it doesn't reach the table it will show you that there is a mirror there because you will see its reflection (*Fig. 98*).

It is quite enough to have a mirror stretching from one table-leg to the other to give one the illusion that there is nothing beneath the table—provided, of course, that the mirror doesn't reflect the furnishings of the room or the audience. That is why it is absolutely necessary for the room to be bare and its walls all alike. The floor too should be in one tone, devoid of all ornamental design, and the audience must be kept at a respectful distance. As you see, the "secret" is as simple as pie, but until you're in the know, you just gape.

Fig. 98. The secret of the lopped-off head

Sometimes the trick is still fancier. First the conjuror shows you a bare table, with nothing on top or beneath it. Then a closed box that is supposed to have the "live head" inside, but which is really empty, is brought onto the stage. The conjuror puts the box on the table and opens up the front flap. And lo! a speaking head appears. You've most likely guessed that the upper board of the table has sort of a trap-door in it through which the man squatting under the table behind the mirror pokes his head when the bottomless empty box is placed on the table. There are other ways of doing this trick. You'll probably be able to work it out for yourself.

IN FRONT OR BEHIND

There are many household things which are not used properly. You already know that some don't use ice properly to cool a drink; they place it on top of the ice instead of *beneath* the ice. Nor does everyone know how to use a mirror properly. Quite often one may put a lamp behind oneself to "light up" one's reflection in the mirror instead of throwing the light on one's own person. Since there are many women who do that, I hope the women among my readers will put the lamp in front of themselves when they want to use a mirror.

IS A MIRROR VISIBLE?

There, again, is proof that what we know about the ordinary mirror is not enough, because most answer this question wrongly, even though all use mirrors every day. Those who think that they can see a mirror are mistaken. A good, clean mirror is invisible. You can see its frame, its rim and everything reflected in it, but you'll never see the mirror itself unless it's dirty. In contrast to a *dispersing* surface—one that scatters light in all directions—every *reflecting* surface is invisible. In ordinary practices a reflecting surface is a polished one, and a dispersing surface, a dull one. All tricks and optical illusions using mirrors—the "speaking head", for instance—are based precisely on their invisibility. All that you do see is the reflection in the mirror of different objects.

IN THE LOOKING-GLASS

When we look in the looking-glass we see ourselves, many will say, adding that what we see is the exact copy of our own person down to the minutest detail.

Let's test that statement. Suppose you have a mole on your right cheek. The person you see in the mirror has a mole on his left cheek. You may be brushing your hair to the *right*; your double in the mirror will be doing it to the *left*. Your right brow may be a bit higher and thicker than your left one; with your copy in the mirror it's the other way round. You keep your watch in your right waistcoat pocket and your wallet in the left pocket; your double has quite opposite habits. Note the

Fig. 99. Use a mirror

dial of his watch. Your watch isn't at all like that. The figures and their arrangement are most unusual. You see an eight marked as it has never been marked before—as IIX—and standing where the twelve ought to be. Meanwhile there is no twelve at all. After a six comes a five, a four and so on. The hands of the watch's double in the mirror move the other way.

To cap it all, he has a physical handicap which you most likely don't have. He's left-handed. He writes, sews and eats with his left hand. And he'll stretch out his left hand

to shake your right one. Then, does he know his letters? At any rate his knowledge is of a most peculiar brand. I greatly doubt whether you will ever be able to read a single line in the book he holds or make out a single word in his left-handed scribble. Such is the person who claims to be your exact copy, the person you claim is exactly like you!

But joking apart, if you really think that by looking in the mirror you are observing yourself, you are mistaken. The face, body and clothing of most people are not strictly symmetrical, but usually we don't notice that. The right side is not quite the same as the left side. In the looking-glass your left side assumes all the peculiar features of your right side and vice versa, so that you actually have a reflection that often produces quite a different impression than you do yourself.

MIRROR DRAWING

The fact that you and your reflection are not totally alike stands out still more when you do the following. Sit down at a table facing an upright mirror. Then take a piece of paper and try to draw, say, a rectangle with intersecting diagonals, by looking at the reflection of your hand. This seemingly simple task becomes incredibly difficult.

Fig. 100. Drawing in front of a looking-glass

As we grow up our visual impressions and motive sensations reach a definite degree of accord. The mirror violates this harmony as it gives us a distorted visual image of our hands in motion. Force of habit cries out against every move you make: you want to draw a line towards the right, but your hand pulls the pencil towards the left. You get still stranger results when you try in this manner to draw still more intricate figures or write something. You are bound to make a most comical mess of things.

The inky imprints on blotting paper are also a mirror-like symmetrical reflection of your handwriting. But try to read them. You won't be able to make out a single word, even when the letters seem quite distinct. The writing will be slanted leftwise and all the strokes are topsy-turvy. However, as soon as you try to read this muddle in a mirror, everything straightens itself out and you recognise your own customary handwriting. Actually, the mirror gives you a symmetrical reflection of what in itself is a symmetrical reflection of your own handwriting.

SHORTEST AND FASTEST

In a homogeneous medium light propagates rectilinearly, that is in the fastest way possible. Light again picks the fastest route when reflecting from a mirror. Let us trace its passage. In *Fig. 101 A* is the source

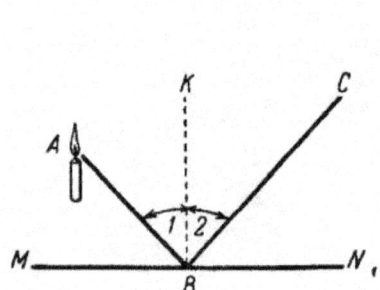 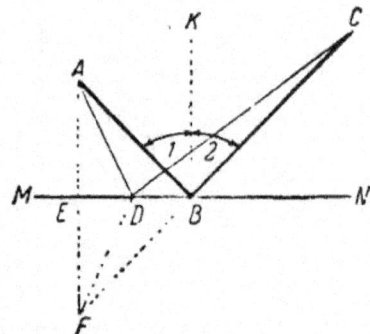

Fig. 101. The angle of reflection 2 is equal to the angle of incidence 1

Fig. 102. Reflecting light chooses the shortest path

of light, a candle, *MN*—a mirror, and *ABC*—the ray's passage from *A* to the eye *C*. The straight line *KB* is perpendicular to *MN*.

According to the laws of optics, the angle of reflection 2 is equal to the angle of incidence 1. Once we know this, we can easily prove that of all possible routes from *A* to *C*, that bounce off the mirror *MN*, *ABC* is the shortest. To prove that this is so, let us compare *ABC* with some other route—for example, *ADC* (*Fig. 102*). Drop the perpendicular *AE* from point *A* onto *MN* and continue it further until it intersects with the continuation of the ray *BC* at point *F*. Then join points *F* and *D* by a

138

straight line. Now let us see first whether the two triangles ABE and EBF are equal. They are both right triangles and both have the side EB adjacent to the right angle. Besides that, the angles EFB and EAB are equal as they are respectively equal to the angles 2 and 1. Consequently, AE is equal to EF. Hence, the right triangles AED and EDF are equal because their respective sides adjacent to the right angles are equal. Consequently, AD is equal to DF.

We can thus replace the route ABC by the equal CBF route—since AB is equal to FB—and the ADC route by the CDF route. Comparing CBF and CDF, we see that the straight line CBF is shorter than the broken line CDF. Consequently, the ABC route is shorter than the ADC one. Q.E.D.!

Wherever point D may be, the ABC route will always be shorter than the ADC one, provided of course the angle of reflection is equal to the angle of incidence. As we see, light indeed chooses the shortest and fastest of all possible routes between its source, the mirror, and the eye. This was first pointed out by Hero of Alexandria, that celebrated 3rd-century Greek mathematician.

AS THE CROW FLIES

The ability to find the shortest way in cases like the one we discussed may come in handy when solving some brain-teasers. Take the following case.

Fig. 103. The problem of the crow. Find the shortest line of flight to the ground and to the fence

Fig. 104. The solution of the problem of the crow

A crow is perched on a branch, and there are some grains of millet scattered on the ground below. The crow swoops down, pecks at the millet and then flies up to perch on the fence. The question is: Where should the crow peck in order to take the shortest possible route? (*Fig. 103.*) This is an absolutely similar problem to the one just discussed. So we can easily supply the right answer: the crow should follow the path of the ray of light. In other words, it should fly so that angle *1* is equal to angle *2* (*Fig. 104*), which, as we already know, is the shortest way possible.

THE KALEIDOSCOPE

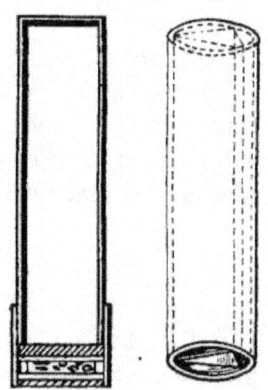

Fig. 105. A kaleidoscope

I suppose you all know what the kaleidoscope is. This amusing toy has a handful of various coloured bits of glass which are placed between two or three flat mirrors. They form extremely beautiful figures which change symmetrically with the slightest twist of the kaleidoscope. Though a very common toy, few suspect the tremendous assortment of different patterns 'one can get. Imagine that you have a kaleidoscope with 20 bits of glass inside and turn it to get ten new patterns every minute. How much time would you need to see all the patterns these 20 bits of glass could form? Even the wildest of imaginations would never provide the right answer. The oceans would dry and the mountains crumble before you saw all; you would need at least 500,000 million years to see every figure produced!

The infinitely different and eternally changing patterns that this toy provides have long intrigued art designers, whose combined imaginations will never match the inexhaustible ingenuity with which it suggests lovely ornamental motifs for wallpaper, carpets and other fabrics. But among the general public it no longer excites the interest it did a hundred years ago when it was a fascinating novelty and when poets composed odes in its honour.

The kaleidoscope was invented in England in 1816. Some twelve to eighteen months later it was already arousing universal admiration. In the July 1818 issue of the Russian magazine *Blagonamerenni (Loyal)*, the fabulist A. Izmailov wrote about it: "Neither poetry nor prose can describe all that the kaleidoscope shows you. The figures change with every twist, with no new one alike. What beautiful patterns! How wonderful for embroidering! But where would one find such bright silks? Certainly a most pleasant relief from idle boredom—much better than to play patience at cards.

"They say that the kaleidoscope was known way back in the 17th century. At any rate, some time ago it was revived and perfected in England to cross the Channel a couple of months ago. One rich Frenchman ordered a kaleidoscope for 20,000 francs, with pearls and gems instead of coloured bits of glass and beads."

Izmailov then provides an amusing anecdote about the kaleidoscope and finally concludes on a melancholic note, extremely characteristic of that backward time of serfdom: "The imperial mechanic Rospini, who is famed for his excellent optical instruments, makes kaleidoscopes which he sells for 20 rubles a piece. Doubtlessly, far more people will want them than to attend the lectures on physics and chemistry from which—to our regret and surprise—that loyal gentleman, Mr. Rospini, has derived no profit."

For long the kaleidoscope was nothing more than an amusing toy. Today it is used in pattern designing. A device has been invented to photograph the kaleidoscope figures and thus mechanically provide sundry ornamental patterns.

PALACES OF ILLUSIONS AND MIRAGES

I wonder what sort of a sensation we would experience if we became midgets the size of the bits of glass and slipped into the kaleidoscope? Those who visited the Paris World Fair in 1900 had this wonderful opportunity. The so-called "Palace of Illusions" was a major attraction there—a place very much like the insides of a huge rigid kaleidoscope. Imagine a hexagonal hall, in which each of the six walls was a large, beautifully polished mirror. In each corner it had architectural embellish-

ments—columns and cornices—which merged with the sculptural
adornments of the ceiling. The visitor thought he was one of a teeming
crowd of people, looking all alike, and filling an endless enfilade of
columned halls that stretched on every side as far as the eye could
see. The halls shaded horizontally in *Fig. 106* are the result of a single
reflection, the next twelve, shaded perpendicularly, the result of a
double reflection, and the next eighteen, shaded slantwise, the result
of a triple reflection. The halls multiply in number with each new mul-

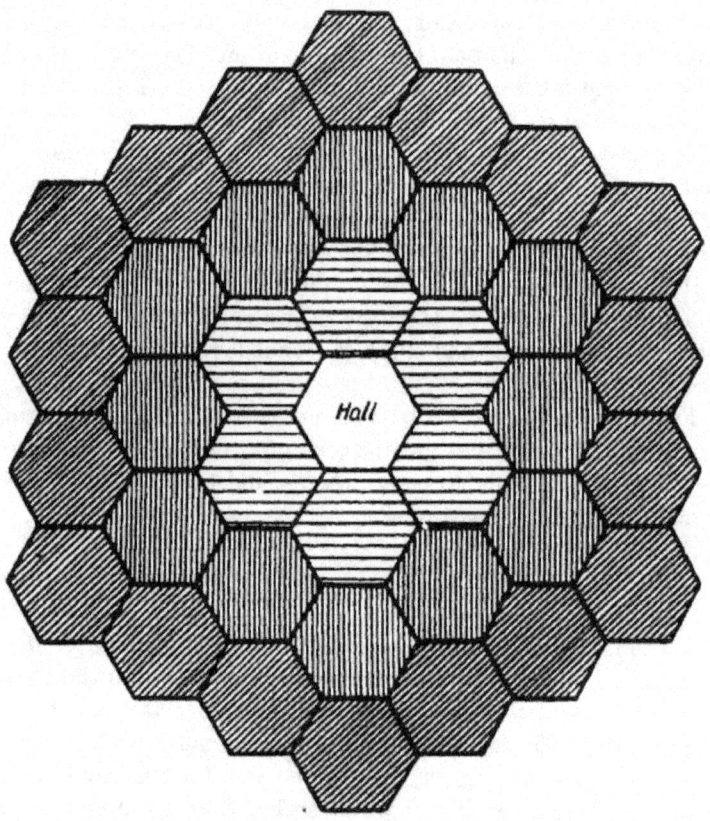

Fig. 106. A three-fold reflection from the walls of the central
hall produces 36 halls

tiple reflection, depending, naturally, on how perfect the mirrors are and whether they are disposed at exact parallels. Actually, one could see only 468 halls—the result of the 12th reflection.

Everybody familiar with the laws that govern the reflection of light will realise how the illusion is produced. Since we have here three pairs of parallel mirrors and ten pairs of mirrors set at angles to each other, no wonder they give so many reflections.

The optical illusions produced by the so-called Palace of Mirages at the same Paris Exposition were still more curious. Here the endless reflections were coupled with a quick change in decorations. In other words, it was a huge but seemingly movable kaleidoscope, with the spectators inside. This was achieved by introducing in the hall of mirrors hinged revolving corners—much in the manner of a revolving stage. *Fig. 107* shows that three changes, corresponding to the corners *1*, *2* and *3*, can be effected. Supposing that the first six corners are decorated as a tropical

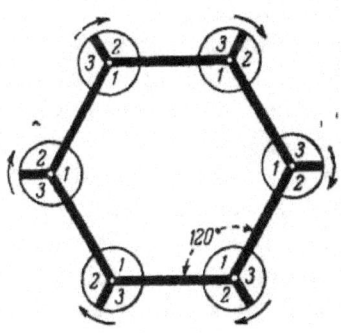

Fig. 107

Fig. 108. The secret of the "Palace of Mirages"

forest, the next six corners as the interior of a sheikh's palace, and the last six as an Indian temple. One turn of the concealed mechanism would be enough to change a tropical forest into a temple or palace. The entire trick is based on such a simple physical phenomenon as light reflection.

WHY LIGHT REFRACTS AND HOW

Many think the fact that light refracts when passing from medium to medium is one of Nature's whims. They simply can't understand why

Fig. 109. Refraction of light explained

light does not keep on in the same direction as before but has to strike out obliquely. Do you think so too? Then you'll probably be delighted to learn that light behaves just as a marching column of soldiers does when they step from a paved road to one full of ruts.

Here is a very simple and instructive illustration to show how light refracts. Fold your tablecloth and lay it on the table as shown in *Fig. 109*. Incline the table-top slightly. Then set a couple of wheels on one axle—from a broken toy steam engine or some other toy—rolling down it. When its path is set at right angles to the tablecloth fold there is no refraction, illustrating the optical law, according to which light falling perpendicularly on the boundary between two different media does not bend. But when its path is set obliquely to the tablecloth fold the direction changes at this point—the boundary between two different media, in which we have a change in velocity.

When passing from that part of the table where velocity is greater (the uncovered part) to that part where velocity is less (the covered part), the direction ("the ray") is nearer to the "normal incidence". When rolling the other way the direction is farther away from the normal.

This, incidentally, explains the substance of refraction as due to the change in light velocity in the new medium. The greater this change is, the wider the angle of refraction is, since the "refractive index", which shows how greatly the direction changes, is nothing but the ratio of the two velocities. If the refractive index in passing from air to water is 4/3, it means that light travels through the air roughly 1.3 times faster than through water. This leads us to another instructive aspect of light propagation. Whereas, when reflecting, light follows the *shortest* route, when refracting, it chooses the *fastest* way; no other route will bring it to its "destination" sooner than this crooked road.

LONGER WAY FASTER

Can a crooked route really bring us sooner to our destination than the straight one? Yes—when we move with different speeds along different sections of our route. Villagers living between two railway stations A and B, but closer to A, prefer to walk or cycle to station A and board the train there for station B, if they want to get to station B *faster*, than to take the *shorter* way which is straight to station B.

Another instance. A cavalry messenger is sent with despatches from point A to the command post at point C (*Fig. 110*). Between him and the command post lie a strip of turf and a strip of soft sand, divided by the straight line EF. We know that it takes twice the time to cross sand than it does to cross turf. Which route would the messenger choose to deliver the despatches sooner?

At first glance one might think it to be the straight line between A and C. But I don't think a single

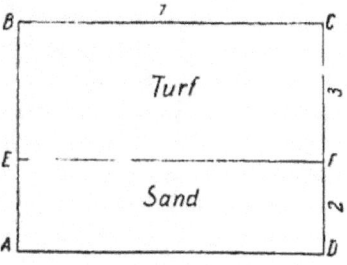

Fig. 110. The problem of the cavalry messenger. Find the fastest way from A to C

horseman would pick that route. After all, since it takes a longer time
to cross sand, a cavalryman would rightly think it better to cut the time
spent by crossing the sand less obliquely. This would naturally length-
en his way across the turf. But since the horse would take him
across it twice as fast, this longer distance would actually mean less
time spent. In other words, the horseman should follow a road that
would *refract* on the boundary between sand and turf, moreover, with
the path across the turf forming a wider angle with the perpendicular
to this boundary than the path across the sand.

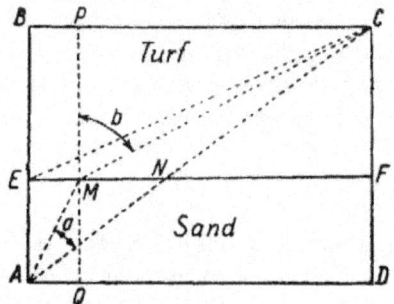

Fig. 111. The problem of the cavalry
messenger and its solution. The fastest way is *AMC*

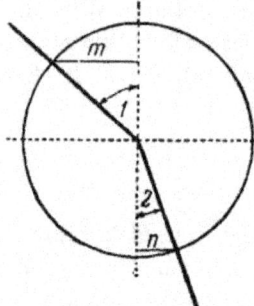

Fig. 112. What is the sine? The relation of *m* to the radius is the sine
of angle *1*, while the relation of *n*
to the radius is the sine of angle *2*

Anyone will realise that the straight path *AC* is actually not the quickest way and that considering the different width of the two strips and
the distances as given in *Fig. 110*, the messenger will reach his destination sooner if he takes the crooked road *AEC* (*Fig. 111*). *Fig. 110* gives
us a strip of sand two kilometres wide, and a strip of turf three kilometres wide. The distance *BC* is seven kilometres. According to Pythagoras,
the entire route from *A* to *C* (*Fig. 111*) is equal to $\sqrt{5^2 + 7^2} = \sqrt{74} =$
$= 8.6$ km. Section *AN*—across the sand—is two-fifths of this, or 3.44 km.
Since it takes twice as long to cross sand than it does to cross turf, the
3.44 km of sand mean from the time angle 6.88 km of turf. Hence the
8.6 km straight-line route *AC* is equivalent to 12.04 km across turf. Let
us now reduce to "turf" the crooked *AEC* route. Section *AE* is two kilo-

146

metres, which corresponds to four kilometres in time across turf. Section EC is equal to $\sqrt{3^2 + 7^2} = \sqrt{58} = 7.6$ km, which, added to four kilometres, results in a total of 11.6 km for the crooked AEC route.

As you see, the "short" straight road is 12 km across turf, while the "long" crooked road only 11.6 km across turf, which thus saves 12.00—11.60=0.40 km, or nearly half a kilometre. But this is still not the *quickest* way. This, according to theory, is that—we shall have to invoke trigonometry—in which the ratio of the sine of angle b to the sine of angle a is the same as the ratio of the velocity across turf to that across sand, i. e., a ratio of 2:1. In other words, we must pick a direction along which the sine of angle b would be twice the sine of angle a. Accordingly, we must cross the boundary between the sand and turf at point M, which is one kilometre away from point E. Then sine $b = \dfrac{6}{\sqrt{3^2+6^2}}$, while sine $a = \dfrac{1}{\sqrt{1+2^2}}$, and the ratio of $\dfrac{\sin b}{\sin a} = \dfrac{6}{\sqrt{45}} : \dfrac{1}{\sqrt{5}} = \dfrac{6}{3\sqrt{5}} : \dfrac{1}{\sqrt{5}} = 2$, which is exactly the ratio of the two velocities. What would this route, reduced to "turf", be? $AM = \sqrt{2^2 + 1^2} = 4.47$ km across turf. $MC = \sqrt{3^2 + 6^2} = 6.49$ km. This adds up to 10.96 km, which is 1.08 km *shorter* than the straight road of 12.04 km across turf.

This instance illustrates the advantage to be derived in such circumstances by choosing a crooked road. Light naturally takes this fastest route because the law of light refraction strictly conforms to the proper mathematical solution. The ratio of the sine of the angle of refraction to the sine of the angle of incidence is the same as the ratio of the velocity of light propagation in the new medium to that in the old medium; this ratio is the refractive index for the specified media. Wedding the specific features of reflection and refraction we arrive at the "Fermat principle"—or the "principle of least time" as physicists sometimes call it—according to which light *always takes the fastest route.*

When the medium is heterogeneous and its refractive properties change gradually—as in our atmosphere, for instance—again "the principle of least time" holds. This explains the slight curvature in light as it comes from the celestial objects through our atmosphere. Astronomers call this "atmospheric refraction". In our atmosphere, which becomes denser and denser the closer we get to the ground, light bends in such a way that the inside of the bend faces the earth. It spends more time in higher atmospheric layers, where there is less to retard its progress, and less time in the "slower" lower layers, thus reaching its destination more quickly than were it to keep to a strictly rectilinear course.

The Fermat principle applies not only to light. *Sound* and all *waves* in general, whatever their nature, travel in accord with this principle. Since you probably want to know why, let me quote from a paper which the eminent physicist Schrödinger read in 1933 in Stockholm when receiving the Nobel Prize. Speaking of how light travels through a medium with a gradually changing density, he said:

"Let the soldiers each firmly grasp one long stick to keep strict breast-line formation. Then the command rings out: Double! Quick! If the ground gradually changes, first the right end, and then the left end will move faster, and the breast-line will swing round. Note that the route covered is not straight but crooked. That it strictly conforms to the shortest, as far as the time of arrival at the destination over this particular ground is concerned, is quite clear, as each soldier tried to run as fast as he could."

THE NEW CRUSOES

If you have read Jules Verne's *Mysterious Island*, you might remember how its heroes, when stranded on a desert isle, lit a fire though they had no matches and no flint, steel and tinder. It was lightning that helped Defoe's Robinson Crusoe; by pure accident it struck a tree and set fire to it. But in Jules Verne's novel it was the resourcefulness of an educated engineer and his knowledge of physics that stood the heroes in good stead. Do you remember how amazed that naïve sailor Pencroft was when, coming back from a hunting trip, he found the engineer and the reporter seated before a blazing bonfire?

"'But who lighted it?' asked Pencroft.

"'The sun!'

"Gideon Spilett was quite right in his reply. It was the sun that had furnished the heat which so astonished Pencroft. The sailor could scarcely believe his eyes, and he was so amazed that he did not think of questioning the engineer.

"'Had you a burning-glass, sir?' asked Herbert of Harding.

"'No, my boy,' replied he, 'but I made one.'

"And he showed the apparatus which served for a burning-glass. It was simply two glasses which he had taken off his own and the reporter's watch. Having filled them with water and rendered their edges adhesive by means of a little clay, he thus fabricated a regular burning-glass, which, concentrating the solar rays on some very dry moss, soon caused it to blaze."

I dare say you would like to know why the space between the two watch glasses had to be filled with water. After all, wouldn't an air-filling focus the sun's rays well enough? Not at all. A watch glass is bounded by two—outer and inner—parallel (concentric) surfaces. Physics tells us that when light passes through a medium bounded by such surfaces it hardly changes its direction at all. Nor does it bend when passing through the second watch glass. Consequently, the rays of light cannot be focussed on one point. To do this we must fill up the empty space between the glasses with a transparent substance that would refract rays better than air does. And that is what Jules Verne's engineer did.

Any ordinary ball-shaped water-filled carafe will act as a burning-glass. The ancients knew that and also noticed that the water didn't warm up in the process. There have been cases when a carafe of water inadvertently left to stand in the sunlight on the sill of an open window set curtains and tablecloths on fire and charred tables. The big spheres of coloured water, which were traditionally used to adorn the show-windows of chemist's shops, now and again caused fires by igniting the inflammable substances stored nearby.

A small round retort—12 cm in diameter is quite enough—full of water will do to boil water in a watch glass. With a focal distance of 15 cm (the focus is very close to the retort), you can produce a temperature of 120° C. You can light a cigarette with it just as easily as with a glass. One must note, however, that a glass lens is much more effective than a water-filled one, firstly, because the refractive index of water is much less, and, secondly, because water intensively absorbs the infra-red rays which are so very essential for heating bodies.

It is curious to note that the ancient Greeks were aware of the ignition effect of glass lenses a thousand odd years before eyeglasses and spyglasses were invented. Aristophanes speaks of it in his famous comedy *The Cloud*. Socrates propounds the following] problem to Streptiadis:

"Were one to write a promissory note on you for five talents, how would you destroy it?

"*Streptiadis*: I have found a way which you yourself will admit to be very artful. I suppose you have seen the wondrous, transparent stone that burns and is sold at the chemist's?

"*Socrates*: The burning-glass, you mean?

"*Streptiadis*: That is right.

"*Socrates*: Well, and what?

"*Streptiadis*: While the notary is writing I shall stand behind him and focus the sun on the promissory note and melt all he writes."

I might explain that in Aristophanes's days the Greeks used to write on waxed tablets which easily melted.

ICE HELPS TO LIGHT FIRE

Even ice, provided it is transparent enough, can serve as a convex lens and consequently as a burning-glass. Let] me note, furthermore, that in this process the ice does not warm up and melt. Its refractive index is a wee bit less than that of water, and since a spherical water-filled vessel can be used as a burning-glass, so can a similarly shaped lump of ice. An ice "burning-glass" enabled Dr. Clawbonny in Jules Verne's *The Adventures of Captain Hatteras* to light a fire when the travellers found themselves stranded without a fire or anything to light it in terribly cold weather, with the mercury at 48° C below zero.

"'This is terrible ill-luck,' the captain said.

"'Yes,' replied the doctor.

"'We haven't even a spyglass to make a fire with!'

"'That's a great pity,' the doctor remarked, 'because the sun is strong enough to light tinder.'

"'We'll have to eat the bear raw, then,' said the captain.

"'As a last resort, yes,' the doctor pensively replied. 'But why not....'

"'What?' Hatteras inquired.

"'I've got an idea.'

"'Then we're saved,' exclaimed the bosun.

"'But...' the doctor was hesitant.

"'What is it?' asked the captain.

"'We haven't got a burning-glass, but we can make one.'

"'How?' asked the bosun.

"'From a piece of ice!'

"'And you think....'

"'Why not? We must focus the sun's rays on the tinder and a piece of ice can do that. Fresh-water ice is better though—it's more transparent and less liable to break.'

Fig. 113. "The doctor focussed the sun's bright rays on the tinder"

"'The ice boulder over there,' the bosun pointed to a boulder some hundred steps away, 'seems to be what we need.'

"'Yes. Take your axe and let's go.'

"The three walked over to the boulder and found that it was indeed of fresh-water ice.

"The doctor told the bosun to chop off a chunk of about a foot in diameter, and then he ground it down with his axe, his knife, and finally polished it with his hand and produced a very good, transparent burning-glass. The doctor focussed the sun's bright rays on the tinder which began to blaze a few seconds later."

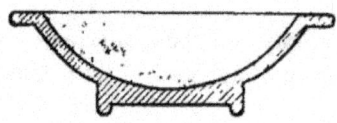

Fig. 114. A bowl for making an ice burning-glass

Jules Verne's story is not an impossibility. The first time this was ever done with success was in England in 1763. Since then ice has been used more than once for the purpose. It is, of course, hard to believe that one could make an ice burning-glass with such crude tools as an axe and knife and "one's hand" in a frost of 48° C below zero. There is, however, a much simpler way: pour some water into a bowl of the proper shape, freeze it, and then take out the ice by slightly heating the bottom of the bowl. Such a "burning-glass" will work only in the open air on a clear and frosty day. Inside a room behind closed windows it is out of the question, because the glass panes absorb much of the solar energy and what is left of it is not strong enough.

HELPING SUNLIGHT

Here is one more experiment which you can easily do in wintertime. Take two pieces of cloth of the same size, one black and the other white, and put them on the snow out in the sun. An hour or two later you will find the black piece half-sunk, while the white piece is still where it was. The snow melts sooner under the black piece because cloth of this colour absorbs most of the solar rays falling on it, while white cloth disperses most of the solar rays and consequently warms up much less.

This very instructive experiment was first performed by Benjamin

Franklin, the American scientist of War for Independence fame, who won immortality for his invention of the lightning conductor.

"I took a number of little square pieces of broad cloth from a tailor's pattern card, of various colours. There were black, deep blue, lighter blue, green, purple, red. yellow, white, and other colours, or shades of colours. I laid them all out upon the snow in a bright sunshiny morning. In a few hours (I cannot now be exact as to the time), the black, being warmed most by the sun, was sunk so low as to be below the stroke of the sun's rays; the dark blue almost as low, the lighter blue not quite so much as the dark, the other colours less as they were lighter; and the quite white remained on the surface of the snow, not having entered it at all.

"What signifies philosophy that does not apply to some use?—May we not learn from hence, that black clothes are not so fit to wear in a hot sunny climate or season, as white ones; because in such clothes the body is more heated by the sun when we walk abroad, and we are at the same time heated by the exercise, which double heat is apt to bring on putrid dangerous fevers?... That summer hats for men or women' should be white, as repelling that heat which gives headaches to many, and to some the fatal stroke that the French call the *coup de soleil*?... That fruit walls being blacked may receive so much heat from the sun in the daytime, as to continue warm in some degree through the night, and thereby preserve the fruit from frosts, or forward its growth?—with sundry other particulars of less or greater importance, that will occur from time to time to attentive minds?"

The benefit that can be drawn from this knowledge was well illustrated during the expedition to the South Pole that the Germans made aboard the good ship *Hauss* in 1903. The ship was jammed by ice-packs and all methods usually applied in such circumstances—explosives and ice-saws—proved abortive. Solar rays were then invoked. A two-kilometre long strip, a dozen metres in width, of dark ash and coal was strewn from the ship's bow to the nearest rift. Since this happened during the Antarctic summer, with its long and clear days, the sun was able to accomplish what dynamite and saws had failed to do. The ice melted and cracked all along the strip, releasing the ship from its clutches.

MIRAGES

I suppose you all know what causes a mirage. The blazing sun heats up the desert sands and lends to them the property of a mirror because the density of the hot surface layer of air is less than the strata higher up. Oblique rays of light from a remote object meet this layer of air and curve upwards from the ground as if reflected by a mirror after striking it at a very obtuse angle. The desert-traveller thus thinks he is seeing a sheet of water which reflects the objects standing on its banks (*Fig. 115*).

Fig. 115. Desert mirages explained. This drawing, usually given in textbooks, shows too steeply the ray's course towards the ground

Rather should we say that the hot surface layer of air reflects not like a mirror but like the surface of water when viewed from a submarine. This is not an ordinary reflection but what physicists call total reflection, which occurs when light enters the layer of air at an extremely obtuse angle, far greater than the one in the figure. Otherwise the "critical angle" of incidence will not be exceeded.

Please note—to avoid misunderstanding—that a denser strata must be above the rarer layers. However, we know that denser air is heavier and always seeks to descend to take the place of lighter lower layers and force them upwards. Why, in the case of a mirage, is the denser air above the rarer air? Because air is in constant motion. The heated surface air keeps on being forced up by a new replacing lot of heated air. This is responsible for some rarefied air always remaining just above the hot sand. It need not be the same rarefied air all the time—but that is something that makes no difference to the rays.

This phenomenon has been known from times immemorial. (A somewhat different mirage appearing in the air at a higher level than the observer is caused by reflection in upper rarefied layers.) Most people think this classical type of mirage can be observed only in the blazing southern deserts and never in more northerly latitudes. They are wrong. This is frequently to be observed in summer on asphalted roads which, because they are dark, are greatly heated by the sun. The dull road's surface seems to look like a pool of water able to reflect distant objects. *Fig. 116* shows the path light takes in this case. A sufficiently observant person will see these mirages oftener than one might think.

There is one more type of mirage—a side one—which people usually do not have the faintest suspicion about. This mirage, which has been

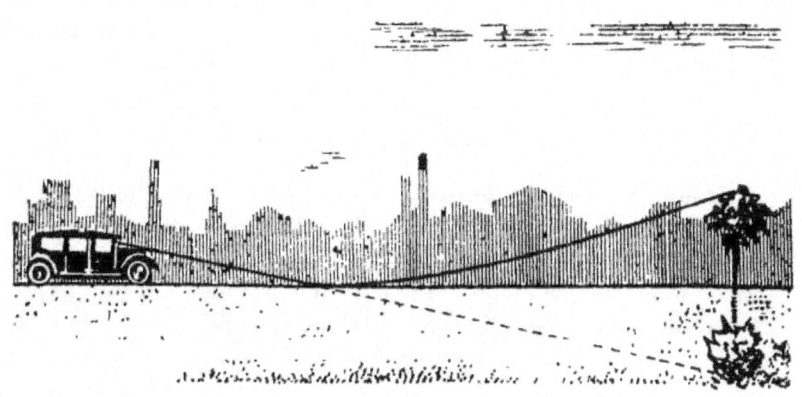

Fig. 116. Mirage on paved highway

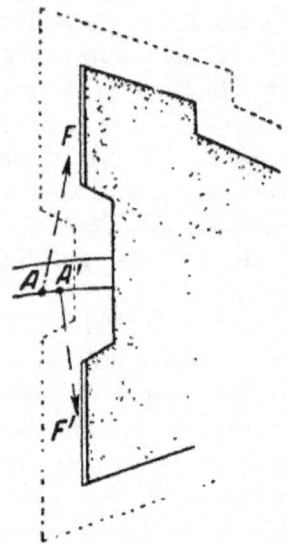

described by a Frenchman, was produced
by reflection from a heated sheer wall. As
he drew near to the wall of a fortress he noticed it suddenly glisten like a polished
mirror and reflect the surrounding land-
scape. Taking a few steps he saw a similar
change in another wall. He concluded that
this was due to the walls having heated up
considerably under the blazing sun. *Fig. 117*
gives the position of the walls (*F* and *F'*)
and the spots (*A* and *A'*) where the observer stood.

The Frenchman found that the mirage re-
curred every time the wall was hot enough
and even managed to photograph the phenomenon.

Fig 118 depicts, on the left, the fortress
wall *F*, which suddenly turned into the glistening mirror on the right, as photographed
from point *A'*. The ordinary grey concrete
wall on the left naturally cannot reflect the
two soldiers near it. But the same wall, miraculously transformed into
a mirror on the right, does *symmetrically* reflect the closer of the two
soldiers. Of course it isn't the wall itself that reflects him, but its surface
layer of hot air. If on a hot summer day you pay notice to walls of
big buildings, you might spot a mirage of this kind.

"THE GREEN RAY"

"Have you ever seen the sun dip into the horizon at sea? No doubt,
you have. Have you ever watched it until the upper rim touches the
horizon and then disappears? Probably you have. But have you ever
noticed what happens on the instant when our brilliant luminary sheds
its last ray—provided the sky is a cloudless, pellucid blue? Probably
not. Don't miss this opportunity. You will see, instead of a red ray, one
of an exquisite green that no artist could ever reproduce and that nature

Fig. 118. Rough, grey wall (left) suddenly seems to act like a polished mirror (right)

herself never displays either in the variously tinted plants or in the most transparent of seas."

This note published in an English newspaper sent the young heroine of Jules Verne's *The Green Ray* in raptures and made her roam the world solely to see this phenomenon with her own eyes. Though, according to Jules Verne, this Scottish girl failed to see the lovely work of nature, still it exists It is no myth, though many legends are associated with it. Any lover of nature can admire it, provided he takes the pains to hunt for it.

Where does the green ray or flash come from? Recall what you saw when you looked at something through a prism. Try the following. Hold the prism at eye level with its broad horizontal plane turned downwards and look through it at a piece of paper tacked to the wall You will see the sheet firstly loom and secondly display a violet-blue rim at the top and a yellow-red edge at the bottom. The elevation is due to refraction, while the coloured rims owe their origin to the property of glass to refract differently light of different colours. It bends violets and blues more than any other colour. That is why we see a violet-blue rim on top. Meanwhile, since it bends reds least, the bottom edge is precisely of this colour.

So that you comprehend my further explanations more easily, I must say something about the origin of these coloured rims. A prism breaks up the white light emitted by the paper into all the colours of the spectrum, giving many coloured images of the paper, disposed in the order of their refraction and often superimposed, one on the other. The combined effect of these superimposed coloured images produces white light (the composition of the spectral colours) but with coloured fringes at top and bottom. The famous poet Goethe who performed this experiment but failed to grasp its real meaning thought that he had debunked Newton's colour theory. Later he wrote his own *Theory of Colours* which is based almost entirely on misconceptions. But I suppose you won't repeat his blunder and expect the prism to colour everything anew.

We see the earth's atmosphere as a vast prism of air, with its base facing us. Looking at the sun on the horizon we see it through a prism of gas. The solar disc has a blue-green fringe on top and a yellow-red one at the bottom. While the sun is above the horizon, its disc's brilliant colour outshines all other less bright bands of colour and we don't see them at all. But during the sunrises and sunsets, when practically the entire disc of the sun is below the horizon, we may spot the blue double-tinted fringe on the upper rim, with an azure blue right on top and a paler blue—produced by the mixing of green and blue—below it. When the air near the horizon is clear and translucent, we see a blue fringe, or the "blue ray". But often the atmosphere disperses the blues and we see only the remaining green fringe—the "green ray". However, most often a turbid atmosphere disperses both blues and greens and then we see no fringe at all, the setting sun assuming a crimson red.

The Pulkovo astronomer G.A. Tikhov, who devoted a special monograph to the "green ray", gives us some tokens by which we may see it. "When the setting sun is crimson-hued and it doesn't hurt to look at it with the naked eye you may be sure that there will be no green flash." This is clear enough: the fact of a red sun means that the atmosphere

158

intensively disperses blues and greens, or, in other words, the whole of the upper rim of the solar disc. "On the other hand," he continues, "when the setting sun scarcely changes its customary whitish yellow and is very bright [in other words, when atmospheric absorption of light is insignificant—Y.P.]—you may quite likely expect the green flash. However, it is important for the horizon to be a distinct straight line with no uneven relief, forests or buildings. We have all these conditions at sea, which explains why seamen are familiar with the green flash."

To sum up: to see the "green ray", you must observe the sun when setting or rising and when the sky is extremely clear. Since the sky at the horizon in southern climes is much more translucent than in northern latitudes, one is liable to see the "green ray" there much oftener. But neither in our latitudes is it so rare as many think—most likely, I suppose, because of Jules Verne. You will detect the "green ray" sooner or later as long as you look hard enough. This phenomenon has been seen even in a spyglass.

Here is how two Alsatian astronomers describe it:

"During the very last minute before the sun sets, when, consequently, a goodly part of its disc is still to be seen, a green fringe hems the waving but clearly etched outline of the sun's ball. But until the sun sets altogether, it cannot be seen with the naked eye. It will be seen only when the sun disappears completely below the horizon. However, should one use a spyglass with a powerful enough magnification—of roughly 100—one will see the entire phenomenon very well. The green fringe is seen some ten minutes before the sun sets at the latest. It incloses the disc's upper half, while a red fringe hems the lower half. At first the fringe is extremely narrow, encompassing at the outset but a few seconds of an arc. As the sun sets, it grows wider, sometimes reaching as much as half a minute of an arc. Above the green fringe one may often spot similarly green prominences, which, as the sun gradually sinks, seem to slide along its rim up to its apex and sometimes break away entirely to shine independently a few seconds before fading" (*Fig. 119*).

Usually this phenomenon lasts a couple of seconds. In extremely favourable conditions, however, it may last much longer. A case of more than 5 minutes has been registered; this was when the sun was setting

Fig. 119. Protracted observation of the "green ray"; it was seen beyond the mountain range for 5 minutes. Top right-hand corner: the "green ray" as seen in a spyglass. The Sun's disc has a ragged shape. 1. The Sun's blinding glare prevents us from seeing the green fringe with the unaided eye. 2. The "green ray" can be seen with the unaided eye when the Sun has almost completely set

behind a distant mountain and the quickly walking observer saw the green fringe as seemingly sliding down the hill (*Fig. 119*).

The instances recorded when the "green ray" has been observed during a sunrise—that is, when the upper rim of our celestial luminary peeps out above the horizon—are extremely instructive, as they debunk the frequent suggestion that the phenomenon is presumably nothing more than an optical illusion to which the eye succumbs owing to the fatigue caused by looking at the brilliant setting sun. Incidentally, the sun is not the only celestial object that sheds the "green ray". Venus has also produced it when setting. (You will find more about mirages and the green flash in M. Minaert's superb book *Light and Colour in Nature*.)

CHAPTER NINE

VISION

BEFORE PHOTOGRAPHY WAS INVENTED

Photography is so ordinary nowadays that we find it hard to imagine how our forefathers, even in the past century, got along without it. In his *Posthumous Papers of the Pickwick Club* Charles Dickens tells us the amusing story of how British prison officers took a person's likeness some hundred or so years ago. The action takes place in the debtors' prison where Pickwick has been brought. Pickwick is told that he'll have to sit for his portrait.

"'Sitting for my portrait!' said Mr. Pickwick.

"'Having your likeness taken, sir.' replied the stout turnkey. 'We're capital hands at likeness here. Take 'em in no time, and always exact. Walk in, sir, and make yourself at home.'

"Mr. Pickwick complied with the invitation, and sat himself down: when Mr. Weller, who stationed himself at the back of the chair, whispered that the sitting was merely another term for undergoing an inspection by the different turnkeys, in order that they might know prisoners from visitors.

"'Well, Sam,' said Mr. Pickwick. 'Then I wish the artists would come. This is rather a public place.'

"'They won't be long, sir, I des-say,' replied Sam. 'There's a Dutch clock, sir.'

"'So I see,' observed Mr. Pickwick.

"'And a bird-cage, sir,' says Sam. 'Veels within veels, a prison in a prison. Ain't it, sir?'

"As Mr. Weller made this philosophical remark, Mr. Pickwick was aware that his sitting had commenced. The stout turnkey having been relieved from the lock, sat down, and looked at him carelessly, from time to time, while a long thin man who had relieved him, thrust his hands beneath his coat-tails, and planting himself opposite, took a good long view of him. A third, rather surly-looking gentleman: who had apparently been disturbed at his tea, for he was disposing of the last remnant of a crust and butter when he came in: stationed himself close to Mr. Pickwick; and, resting his hands on his hips, inspected him narrowly; while two others mixed with the group, and studied his features with most intent and thoughtful faces. Mr. Pickwick winced a good deal under the operation, and appeared to sit very uneasily in his chair; but he made no remark to anybody while it was being performed, not even to Sam, who reclined upon the back of the chair, reflecting, partly on the situation of his master, and partly on the great satisfaction it would have afforded him to make a fierce assault upon all the turnkeys there assembled, one after the other, if it were lawful and peaceable so to do.

"At length the likeness was completed, and Mr. Pickwick was informed, that he might now proceed into the prison."

Still earlier it was a list of "features" that did for such memorised "portraits". In his *Boris Godunov*, Pushkin tells us how Grigory Otrepyev was described in the tsar's edict: "Of short stature, and broad chest; one arm is shorter than the other; the eyes are blue and hair ginger; a wart on one cheek and another on the forehead." Today we needn't do that; we simply provide a photograph instead.

WHAT MANY DON'T KNOW HOW TO DO

Photography was introduced in Russia in the 1840's, first as daguerreotypes—prints on metal plates that were called so after their inventor, Daguerre. It was a very inconvenient method; one had to pose for quite a long stretch—for as long as fourteen minutes or more. "My grandfather," Prof. B.P. Weinberg, the Leningrad physicist, told me, "had to sit for 40 minutes before the camera to get just one daguerreotype, from which, moreover, no prints could be made."

Still the chance to have one's portrait made without the artist's intervention seemed such a wonderful novelty that it took the general public quite a time to get used to the idea. One old Russian magazine for 1845 contains quite an amusing anecdote on the score:

"Many still cannot believe that the daguerreotype acts by itself. One gentleman came to have his portrait done. The owner [the photographer —Y.P.] begged him to be seated, adjusted the lenses, inserted a plate, glanced at his watch, and retired. While the owner was present, the gentleman sat as if rooted to the spot. But he had barely gone out when the gentleman thought it no longer necessary to sit still; he rose, took a pinch of snuff, examined the camera from every side, put his eye to the lens, shook his head, mumbled, 'How ingenious,' and began to meander up and down the room.

"The owner returned, stopped short in surprise at the doorway, and exclaimed: 'What are you doing? I told you to sit still!'

"'Well, I did. I got up only when you went out.'

"'But that was exactly when you should have sat still.'

"'Why should I sit still for nothing?' the gentleman retorted."

We're certainly not so naïve today.

Still, there are some things about photography that many do not know. Few, incidentally, know how one should *look* at a photograph. Indeed, it's not so simple as one might think, though photography has been in existence for more than a century now and is as common as could be. Nevertheless, even professionals don't look at photographs *in the proper way*.

HOW TO LOOK AT PHOTOGRAPHS

The camera is based on the same optical principle as our eye. Everything projected onto its ground-glass screen depends on the distance between the lens and the object. The camera gives a perspective, which we would get with *one eye*—note that!—were our eye to replace the lens. So, if you want to obtain from a photograph the same visual impression that the photographed object produced, we must, firstly, look at the photograph with *one eye* only, and, secondly, hold it *at the proper distance away*.

After all, when you look at a photograph with *both eyes* the picture you get is flat and not three-dimensional. This is the fault of our own vision. When we look at something solid the image it causes on the retina of either eye is not the same (*Fig. 120*). This is mainly why we see objects in relief. Our brain blends the two different images into one that springs into *relief*—this is the basic principle of the stereoscope. On the other hand, if we are looking at something that is flat—a wall, for instance—both eyes get an identical sensory picture telling our brain that the object we are looking at is really flat.

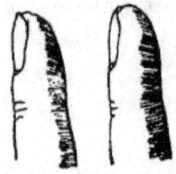

Fig. 120. A finger as !seen separately by the left and right eye when held close to the face ´

Now you should realise the mistake we make when we look at a photograph with both eyes. In this manner we compel ourselves to believe that the picture we have before us is flat. When we look with *both eyes* at a photograph which is really intended only for *one eye*, we prevent ourselves from seeing the picture that the photograph really shows, and thus destroy the illusion which the camera produces with such perfection.

HOW FAR TO HOLD A PHOTOGRAPH

The second rule I mentioned—that of holding the photograph *at the proper distance away* from the eye—is just as important, for otherwise we get the wrong perspective. How far away should we hold a photograph? To recreate the proper picture we must look at the photograph from the same angle of vision from which the camera lens reproduced the image on the ground-glass screen, or in the same way as it "saw" the object being photographed (*Fig. 121*). Consequently, we must hold the photograph at such a distance away from the eye that would be as many times less the distance between the object and the lens as the size of the image on the photograph is less its actual size. In other words,

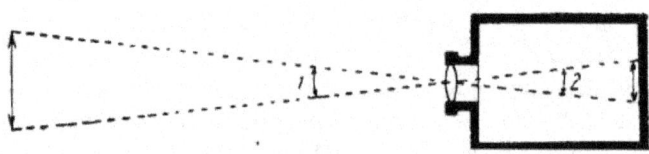

Fig. 121. In a camera angle *1* is equal to angle *2*

we must hold the photograph at a distance which is roughly the same as the focal length of the camera lens.

Since most cameras have a focal length of 12-15 cm (the author has in mind the cameras that were in use when he wrote his *Physics for Entertainment—Ed.*), we shall never be able to get the proper distance for the photographs they give, as the focal length of a normal eye at best (25 cm) is nearly twice the indicated focal length of the camera lens. A photograph tacked on a wall also seems flat because it is looked at from a still greater distance away. Only the short-sighted with their short focal length of vision, as well as children, who are able to accommodate their vision to see objects very close up, will be able to admire the effect that an ordinary photograph produces when we look at it properly with one eye, because when they hold a photograph 12-15 cm away, they get not a flat image but one in relief—the kind of image a stereoscope produces.

I suppose you will now agree with me in noting that it is only due to ignorance that we do not derive the pleasure a photograph can give, and that we often unjustly blame them for being lifeless.

QUEER EFFECT OF MAGNIFYING GLASS

The short-sighted easily see ordinary photographs in relief. What should people with normal eyesight do? Here a magnifying glass will help. By looking at photographs through a magnifying glass with a two-fold power, people with normal eyesight will derive the indicated advantage of the short-sighted, and see them in relief without straining their eyesight.

There is a tremendous difference between the effect thus produced and the impression we get when we look at a photograph with both eyes from quite a distance. It almost amounts to the stereoscopic effect. Now we know why photographs often spring into relief when looked at with one eye through a magnifying glass, which, though a generally known fact, has seldom been properly explained. One reviewer of this book wrote to me in this connection:

"Please take up in a future edition the question of why photographs appear in relief when viewed through a magnifying glass. Because I contend that the involved explanation provided of the stereoscope holds no water at all. Try to look in the stereoscope with one eye. The picture appears in relief despite all that theory has to say."

I am sure you will agree that this does not pick any holes in the theory of stereoscopic vision.

The same principle lies at the root of the curious effect produced by the so-called panoramas, that are sold at toy shops. This is a small box, in which an ordinary photograph—a landscape or a group of people—is placed and viewed through a magnifying glass with one eye, which in itself already gives a stereoscopic effect. The illusion is usually enhanced by some of the objects in the foreground being cut out and placed separately in front of the photograph proper. Our eye is very sensitive to the solidity of objects close by; as far as distant objects are concerned, the impression is much less perceptible.

ENLARGED PHOTOGRAPHS

Can we make photographs so that people with *normal* eyesight are able to see them properly, without using a magnifying glass? We can, merely by using cameras having lenses with a long focal length. You already know that a photograph obtained with the aid of a lens having a focal distance of 25-30 cm will appear in relief when viewed with one eye from the usual distance away.

One can even obtain photographs that won't seem flat even when looked at with *both eyes* from quite a distance. You also know that our brain blends two identical retinal images into one flat picture. However, the greater the distance away from the object, the less our brain is able to do that. Photographs taken with the aid of a lens having a focal distance of 70 cm can be looked at with both eyes without losing the sense of depth.

Since it is incommoding to resort to such lenses, let me suggest another method, which is to *enlarge* the picture you take with any ordinary camera. This increases the distance at which you should look at photographs to get the proper effect. A four- or fivefold enlargement of a photograph taken with a 15 cm lens is already quite enough to obtain the desired effect—you can look at it with both eyes from 60 to 75 centimetres away. True, the picture will be a bit blurred but this is barely discernible at such a distance. Meanwhile, as far as the stereoscopic effect and depth are concerned, you only stand to gain.

BEST SEAT IN MOVIE-HOUSE

Cinema-goers have most likely noticed that some films seem to spring into unusually clear relief—to such an extent at times that one seems to see real scenery and real actors. This depends not on the film, as is often thought, but on where you take your seat. Though motion pictures are taken with cameras having lenses with a very short focal length, their projection on the screen is a hundred times larger—and you can see them with both eyes from quite a distance ($10 \text{ cm} \times 100 = 10 \text{ m}$). The effect of relief is best when you look at the picture from the same angle of vision as the movie camera "looked" when it was shooting the film.

How should one find the distance corresponding to such an optimal angle of vision? Firstly, one must choose a seat *right opposite the middle of the screen*. Secondly, one's seat must be away from the screen at a distance which is as many times the screen's width as the focal length of the movie-camera lens is greater than the width of the film itself. Movie-camera lens usually have a focal length of 35 mm, 50 mm, 75 mm, or 100 mm, depending on the subject being shot. The standard width of film is 24 mm. For a focal length of 75 mm, for instance, we get the proportion:

$$\frac{\text{the distance}}{\text{screen width}} = \frac{\text{focal length}}{\text{film width}} = \frac{75}{24} \approx 3.$$

So, to find how far away you should seat yourself from the screen, you should multiply the width of the screen, or rather the projection onto the screen, by three. If the width is six of your steps, then the best seat would be 18 steps away from the screen. Keep this in mind when trying various devices offering a stereoscopic effect, because one may easily ascribe to the invention what is really due to the tioned.

Reproductions in books and magazines naturally have the same properties as the original photographs from which they were made; they also spring into relief when looked at with one eye from the proper distance. But since different photographs are taken by cameras having lenses with different focal lengths, one can find the proper distance only by trial and error. Cup one eye with your hand and hold the illustration at arm's length. Its plane must be perpendicular to the line of vision and your open eye must be right opposite the middle of the picture. Gradually bring the picture closer, steadily looking at it meanwhile; you easily catch the moment when it appears in clearest relief.

Many illustrations that seem blurred and flat when you look at them in your habitual way acquire depth and clearness when viewed as I suggest. One will even catch the sparkle of water and other such purely stereoscopic effects.

It's amazing that few people know these simple things though they were all explained in popular-science books more than half a century ago. In his *Principles of Mental Physiology, with Their Application to the Training and Discipline of the Mind, and the Study of Its Morbid Conditions*, William Carpenter has the following to say about how one should look at photographs.

"It is remarkable that the effect of this mode of viewing photographic pictures is not limited to bringing out the solid forms of objects; for other features are thus seen in a manner more true to the reality, and therefore more suggestive of it. This may be noticed especially with regard to the representation of *still water*, which is generally one of the most unsatisfactory parts of a photograph; for although, when looked at with *both* eyes, its surface appears opaque, like white wax, a wonderful depth and transparence are often given to it by viewing it with only *one*. And the same holds good also in regard to the characters of *surfaces* from which light is reflected—as bronze or ivory; the material of the object from which the photograph was taken being recognised much more certainly when the picture is looked at with *one* eye, than when *both* are used (unless in stereoscopic combination)."

There is one more thing we must note. Photographic enlargements, as we have seen, are more lifelike; photographs of a reduced size are not. True, the smaller-size photograph gives a better contrast; but it is flat and fails to give the effect of depth and relief. You should now be able to say why: it also reduces the corresponding perspective—which is usually too little as it is.

HOW TO LOOK AT PAINTINGS

All I have said of photographs applies in some measure to paintings as well. They appear best also at the proper distance away, for only then do they spring into relief. It is better, too, to view them with but one eye, especially if they are small.

"It has long been known," Carpenter wrote in the same book, "that if we gaze steadily at a picture, whose perspective projection, lights and shadows, and general arrangement of details, are such as accurately correspond with the reality which it represents, the impression it produces will be much more vivid when we look with *one* eye only, than when we use both; and that the effect will be further heightened, when we carefully shut out the surroundings of the picture, by looking through a tube of appropriate size and shape. This fact has been commonly accounted for in a very erroneous manner. 'We see more exquisitely,' says Lord Bacon, 'with one eye than with both, because the vital spirits thus unite themselves the more and become the stronger'; and other writers, though in different language, agree with Bacon in attributing the result to the *concentration* of the visual power, when only one eye is used. But the fact is, that when we look with *both* eyes at a picture within a moderate distance, we are *forced* to recognise it as a flat surface; whilst, when we look with only *one*, our minds are at liberty to be acted on by the suggestions furnished by the perspective, chiaroscuro, etc.; so that, after we have gazed for a little time, the picture may begin to start into relief, and may even come to possess the solidity of a model."

Reduced photographic reproductions of big paintings often give a greater illusion of relief than the original. This is because the reduced size lessens the ordinarily long distance from which the painting should be looked at, and so the photograph acquires relief, even close up.

THREE DIMENSIONS IN TWO

All I have said about looking at photographs, paintings and drawings, while being true, should not be taken in the sense that there is no other way of looking at flat pictures to get the effect of depth and relief. Every artist, whatever his field—painting, the graphic arts, or photography—strives to produce an impression on the spectator regardless of his "point of view". After all he can't count on everybody viewing his creations with hands cupped over one eye and sizing up the distance for every piece.

Every artist, including the photographer, has an extensive arsenal of means to draw upon to give in two dimensions objects possessing three. The different retinal images produced by distant objects are not the only token of depth. The "aerial perspective" painters employ grading tones and contrasts to make the background blurred and seemingly veiled by diaphanous mist of air, plus their use of linear perspective produces the illusion of depth. A good specialist in art photography will follow the same principles, cleverly choosing lighting, lenses, and also the appropriate brand of photographic paper to produce perspective.

Proper focussing is also very important in photography. If the foreground is sharply contrasted and the remoter objects are "out of focus", this alone is already enough, in many cases, to create the impression of depth. On the contrary, when you reduce the aperture and give both foreground and background in the same contrast, you achieve a flat picture with no depth to it. Generally speaking, the effect a picture produces on the spectator—thanks to which he sees three dimensions in two, irrespective of physiological conditions for visual perception and sometimes in violation of geometrical perspective—depends largely, of course, on the artist's talent.

STEREOSCOPE

Why is it that we see solid objects as things having three dimensions and not two? After all the retinal image is a flat one. So why do we get a sensory picture of geometrical solidity? For several reasons. Firstly, the different lighting of the different parts of objects enables us to perceive their shape. Secondly, the strain we feel when accommodating our eye to get a clear perception of the different distance of the object's different parts also plays a role; this is not a flat picture in which every part of the object depicted is set at the same distance away. And thirdly—the most important cause—is that the two retinal images are different, which is easy enough to demonstrate by looking at some close object, shutting alternately the right and left eye (*Figs. 120* and *122*).

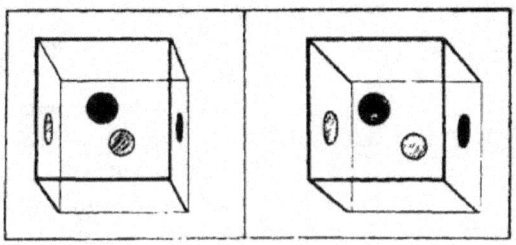

Fig. 122. A spotted glass cube as seen with the
left and right eye

Imagine now two drawings of one and the same object, one as seen by the left eye, and the other, as seen by the right eye. If we look at them so that each eye sees only its "own" drawing, we get instead of two separate flat pictures one in relief. The impression of relief is greater even than the impression produced when we look at a solid object with one eye only.

There is a special device, called the stereoscope, to view these pairs. Older types of stereoscopes used mirrors and the later models convex glass prisms to superimpose the two images. In the prisms—which slightly enlarge the two images, because they are convex—the light coming from the pair is refracted in such a way that its imagined continuation causes this superimposition.

As you see, the stereoscope's basic principle is extremely simple; all the more amazing, therefore, is the effect produced. I suppose most of you have seen various stereoscopic pictures. Some may have used the stereoscope to learn stereometry more easily. However, I shall proceed to tell you about applications of the stereoscope which I presume many of you do not know.

Actually we can—provided we accustom our eyes to it—dispense with the stereoscope to view such pairs, and achieve the same effect, with the sole difference that the image will not be bigger than it usually is in a stereoscope. Wheatstone, the inventor of the stereoscope, made use of this arrangement of nature. Provided here are several stereoscopic drawings—, graded in difficulty—that I would advise you to try viewing without a stereoscope. Remember that you will achieve results

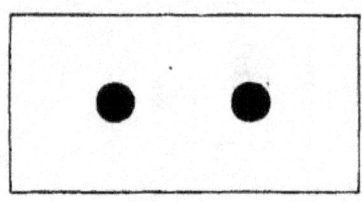

Fig. 123. Stare at the space between the two dots for several seconds. The dots seem to merge

only if you exercise. (Note that not all can see stereoscopically, even in a stereoscope: some—the squint-eyed or people used to working with one eye—are utterly incapable of adjustment to binocular vision; others achieve results only after prolonged exercise. Young people, however, quickly adapt themselves, after a quarter of an hour.)

Start with *Fig. 123* which depicts two black dots. Stare several seconds at the space between them, meanwhile trying to look at an imagined object behind. Soon you will be seeing double, seeing four dots instead of two. Then the two

Fig. 124. Do the same, after which turn to the next exercise

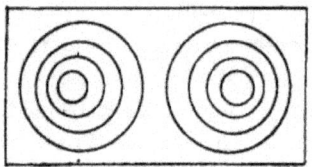

Fig. 125. When these images merge you will see something like the inside of a pipe receding into the distance

extreme dots will swing far apart, while the two innermost dots will close up and become one. Repeat with *Figs. 124* and *125* to see something like the inside of a long pipe receding into the distance.

Then turn to *Fig. 126* to see geometrical bodies seemingly suspended in mid-air. *Fig. 127* will appear as a long corridor or tunnel. *Fig. 128* will produce the illusion of transparent glass in an aquarium. Finally, *Fig. 129* gives you a complete picture, a seascape.

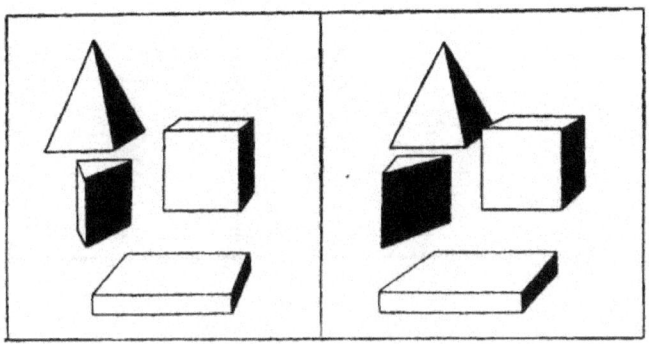

Fig. 126. When these four geometrical bodies merge, they seem to hover in mid-air

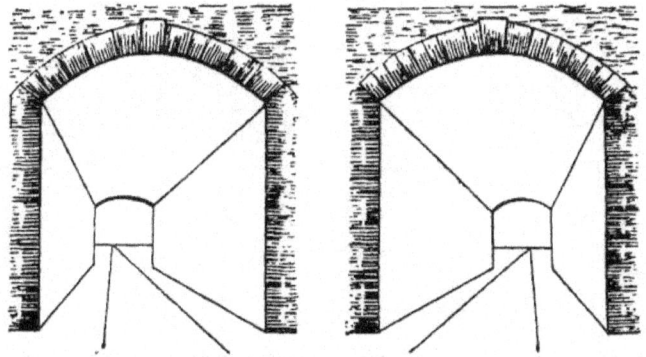

Fig. 127. This pair gives a long corridor receding into the distance

It is easy to achieve results. Most of my friends learned the trick very quickly, after a few tries. The short-sighted and far-sighted needn't take off their glasses; they view the pairs just as they look at any picture. Catch the proper distance at which they should be held by trial and error. See that the lighting is good—this is important.

Now you can try to view stereoscopic pairs in general without a stereoscope. You might try the pairs in *Figs. 130* and *133* first. Don't

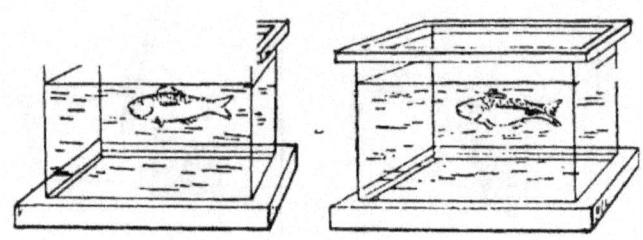

Fig. 128. A fish in an aquarium

Fig. 129. A stereoscopic seascape

overdo this so as not to strain your eyesight. If you fail to acquire the knack, you may use lenses for the far-sighted to make a simple but quite serviceable stereoscope. Mount them side by side in a piece of cardboard so that only their inner rims are available for viewing. Partition off the pairs with a diaphragm.

Fig. 130

Fig. 130 (the upper left-hand corner) gives two photographs of three bottles of presumably one and the same size. However hard you look you cannot detect any difference in size. But there is a difference, and, moreover, a significant one. They seem alike only because they are not set at one and the same distance away from the eye or camera. The bigger bottle is further away than the smaller ones. But which of the three is the bigger bottle? Stare as much as you may, you will never get the answer. But the problem is easily solved by using a stereoscope or exercising binocular vision. Then you clearly see that the left-hand bottle is furthest away, and the right-hand bottle closest. The photo in the upper right-hand corner shows the real size of the bottles.

The stereoscopic pair at the bottom of *Fig. 130* provides a still bigger teaser. Though the vases and candlesticks seem identical there is a great difference in size between them. The left-hand vase is nearly twice as tall as the right-hand one, while the left-hand candlestick, on the contrary, is much smaller than the clock and the right-hand candlestick. Binocular vision immediately reveals the cause. The objects are not in one row; they are placed at different distances, with the bigger objects being further away than the smaller articles. A fine illustration of the great advantage of binocular "two-eyed" vision over "one-eyed" vision!

DETECTING FORGERY

Suppose you have two absolutely identical drawings, of two equal black squares, for instance. In the stereoscope they appear as one square which is exactly alike either of the twin squares. If there is a white dot in the middle of each square, it is bound to show up on the square in the stereoscope. But if you shift the dot on one of the squares slightly off centre, the stereoscope will show one dot—however, it will appear either *in front* of, or *beyond*, the square, not on it. The slightest of differences already produces the impression of depth in the stereoscope. This provides a simple method for revealing forgeries. You need only put the suspected bank-bill next to a genuine one in a stereoscope, to detect the forged one, however cunningly made. The slightest discrepancy, even in one teeny-weeny line, will strike the eye at once—appearing either in front of, or behind, the banknote. (The idea, which was first suggested by Dove in the mid-19th century, is not applicable—for reasons of printing technique—to all currency notes issued today. Still his method will do to distinguish between two proofs of a book-page, when one is printed from newly-composed type.)

AS GIANTS SEE IT

When an object is very far away, more than 450 metres distant, the stereoscopic impression is no longer perceptible. After all the 6 centimetres at which our eyes are set apart are nothing compared with such a distance as 450 metres. No wonder buildings, mountains, and landscapes that are far away seem flat. So do the celestial objects all appear to be at the same distance, though, actually, the moon is much closer than the planets, while the planets, in turn, are very much closer than the fixed stars. Naturally, a stereoscopic pair thus photographed will not produce the illusion of relief in the stereoscope.

There is an easy way out, however. Just photograph distant objects from two points, taking care that they be further apart than our two eyes. The stereoscopic illusion thus produced is one that we would get were our eyes set much further apart than they really are. This is actually how stereoscopic pictures of landscapes are made. They are usually viewed through magnifying (convex) prisms and the effect is most amazing.

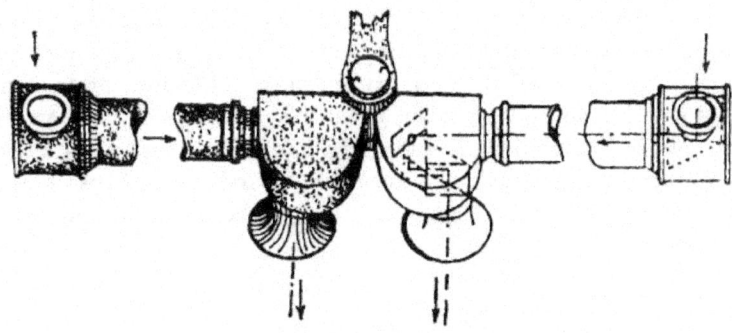

Fig. 131. Telestereoscope

You have probably guessed that we could arrange two spyglasses to present the surrounding scenery in its real relief. This instrument, called a telestereoscope, consists of two telescopes mounted further apart than eyes normally are. The two images are superimposed by means of reflecting prisms (*Fig. 131*).

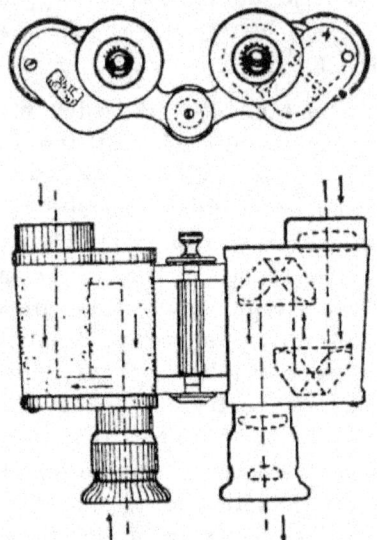

Fig. 132. Prism binoculars

Words fail to convey the sensation one experiences when looking through a telestereoscope, it is so unusual. Nature is transformed; distant mountains spring into relief; trees, rocks, buildings and ships at sea appear in all three dimensions. No longer is everything flat and fixed; the ship, that seems a stationary spot on the horizon in an ordinary spyglass. is moving. That is most likely how the legendary giants saw surrounding nature. When this device has a tenfold power and the distance between its lenses is six times the interocular distance (6.5×6=39 cm), the impression of relief is enhanced 60-fold (6×10), compared with the impression obtained by the naked eye. Even objects 25 kilometres away still appear in discernible relief. For land surveyors, seamen, gunners and travellers this instrument is a godsend, especially if equipped with a range-finder. The Zeiss prism binoculars produces the same effect, as the distance between its lenses is greater than the normal interocular distance (*Fig. 132*). The opera glass, on the contrary, has its lenses set not so far apart, to reduce the illusion of relief— so that the décor and settings present the intended impression.

If we direct our telestereoscope at the moon or any other celestial object we shall fail to obtain any illusion of relief at all. This is only natural, as celestial distances are too big even for such instruments. After all, the 30-50 cm distance between the two lenses is nothing compared with the distance from the earth to the planets. Even if the two telescopes were mounted tens and hundreds of kilometres apart, we would get no results, as the planets are tens of millions of kilometres away.

This is where stereoscopic photography steps in. Suppose we photograph a planet today and take another photograph of it tomorrow. Both photographs will be taken from one and the same point on the globe, but from different points in the solar system, as in the space of 24 hours the earth will have travelled millions of kilometres in orbit. Hence the two photographs won't be identical. In the stereoscope, the pair will produce the illusion of relief. As you see, it is the earth's orbital motion that enables us to obtain stereoscopic photographs of celestial objects. Imagine a giant with a head so huge that its interocular distance ranges into millions of kilometres; this will give you a notion of the unusual effect astronomers achieve by such stereoscopic photography. Stereoscopic photographs of the moon present its mountains in relief so distinct that scientists have even been able to *measure* their height. It seems as if the magic chisel of some super-colossal sculptor has breathed life into the moon's flat and lifeless scenery.

The stereoscope is used today to discover the asteroids which swarm between the orbits of Mars and Jupiter. Not so long ago the astronomer considered it a stroke of good fortune if he was able to spot one of these asteroids. Now it can be done by viewing stereoscopic photographs of this part of space. The stereoscope immediately reveals the asteroid; it "sticks" out.

In the stereoscope we can detect the difference not only in the *position* of celestial objects but also in their *brightness*. This provides the astronomer with a convenient method for tracking down the so-called *variable* stars whose light periodically fluctuates. As soon as a star exhibits a dissimilar brightness the stereoscope detects at once the star possessing that varying light.

Astronomers have also been able to take stereoscopic photographs of the nebulae (Andromeda and Orion). Since the solar system is too small for taking such photographs astronomers availed themselves of our system's displacement amidst the stars. Thanks to this motion in the universe we always see the starry heavens from new points. After the lapse of an interval long enough, this difference may even be detected by the camera. Then we can make a stereoscopic pair, and view it in the stereoscope.

THREE-EYED VISION

Don't think this a slip of the tongue on my part; I really mean three eyes. But how can one see with three eyes? And can one really acquire a third eye?

Science cannot give you or me a third eye, but it can give us the magic power to see an object as it would appear to a three-eyed creature. Let me note first that a one-eyed man can get from stereoscopic photographs that impression of relief which he can't and doesn't get in ordinary life. For this purpose we must project onto a screen in rapid sequence the photographs intended for right and left eyes that a normal person sees with both eyes simultaneously. The net result is the same because a rapid sequence of visual images fuses into one image just as two images seen simultaneously do. (It is quite likely that the surprising "depth" of movie films at times, in addition to the causes mentioned, is due also to this. When the movie camera sways with an even motion—as often happens because of the film-winder—the stills will not be identical and, as they rapidly flit onto the screen will appear to us as one 3-dimensional image.)

In that case couldn't a two-eyed person simultaneously watch a rapid sequence of two photographs with one eye and a third photograph, taken from yet another angle, with the other eye? Or, in other words, a stereoscopic "trio"? We could. One eye would get a single image, but in relief, from a rapidly alternating stereoscopic pair, while the other eye would look at the third photograph. This "three-eyed" vision enhances the relief to the extreme.

The stereoscopic pair in *Fig. 133* depicts polyhedrons, one in white against a black background and the other in black against a white background. How would they appear in a stereoscope? This is what Helmholtz says:

"When you have a certain plane in white on one of a stereoscopic pair and in black on the other, the combined image seems to sparkle,

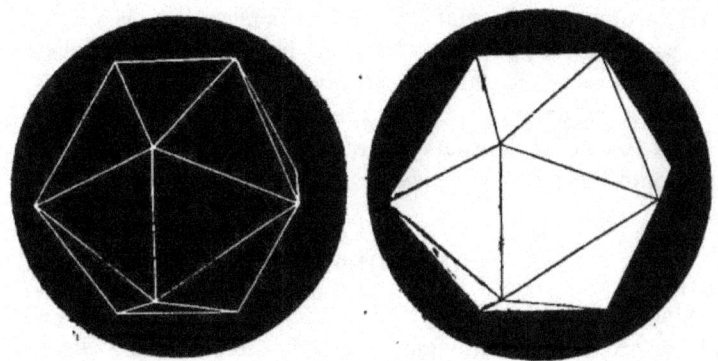

Fig. 133. Stereoscopic sparkle. In the stereoscope this pair produces a sparkling crystal against a black background

even though the paper used for the pictures is dull. Such stereoscopic drawings of models of crystals produce the impression of glittering graphite. The sparkle of water, the glisten of leaves and other such things are still more noticeable in stereoscopic photographs when this is done."

In an old but far from obsolete book, *The Physiology of the Senses. Vision*, which the Russian physiologist Sechenov published in 1867, we find a wonderful explanation of this phenomenon.

"Experiments artificially producing stereoscopic fusion of differently lighted or differently painted surfaces repeat the actual conditions in which we see sparkling objects. Indeed, how does a dull surface differ from a glittering polished one? The first one reflects and diffuses light and so seems identically lighted from every point of observation, while the polished surface reflects light in but one definite direction.

Therefore you can have instances when with one eye you get many re-flected rays, and with the other practically none—these are precisely the conditions that correspond to the stereoscopic fusion of a white surface with a black one. Evidently there are bound to be instances in looking at glistening polished surfaces when reflected light is unevenly distributed between the eyes of the observer. Consequently, the stereoscopic sparkle proves that experience is paramount in the act during which images fuse bodily. The conflict between the fields of vision immediately yields to a firm conception, as soon as the expe-rience-trained apparatus of vision has the chance to attribute the difference to some familiar instance of actual vision."

So the reason we see things *sparkle*—or at least one of the reasons— is that the two retinal images are not the same. Without the stereoscope we would have scarcely guessed it.

TRAIN WINDOW OBSERVATION

I noted a little earlier that different images of one and the same object produce the illusion of relief when in rapid alternation they perceptibly fuse. Does this happen only when we see moving images and stand still ourselves? Or will it also take place when the images are standing still but we are moving? Yes, we get the same illusion, as was only to be expected. Most likely many have noticed that movies shot from an express train spring into unusually clear relief—just as good as in the stereoscope. If we pay heed to our visual perceptions when riding in a fast train or car we shall see this ourselves. Landscapes thus observed spring into clear relief with the foreground distinctly separate from the background. The "stereoscopic radius" of our eyes increases appreciably to far beyond the 450-metre limit of binocular vision for stationary eyes.

Doesn't this explain the pleasant impression we derive from a land-scape when observing it from the window of an express train? Remote objects recede and we distinctly see the vastness of the scenic pano-rama unfolding before us. When we ride through a forest we stereoscop-ically perceive every tree, branch, and leaf; they do not blend into one flat picture as they would to a stationary observer. On a mountain road fast driving again produces the same effect. We seem to sense tangibly the dimensions of the hills and valleys.

One-eyed people will also see this—and I'm sure it will afford a startlingly novel sensation, as this is tantamount to the rapid sequence of pictures producing the illusion of relief, a point mentioned before. (This, incidentally, accounts for the noticeable stereoscopic effect produced by movie films shot from a train taking a bend, when the objects being photographed lie in the radius of this bend. This track "effect" is well-known to cameramen.)

It is as easy as pie to check my statements. Just be mindful of your visual perceptions when riding in a car or a train. You might also notice another amazing circumstance which Dove remarked upon some hundred years ago—what is well forgotten is indeed novel!—that the closer objects flashing by seem smaller in size. The cause has little to do with binocular vision. It's simply because our estimate of distance is wrong. Our subconscious mind suggests that a closer object should really be smaller than usually, to seem as big as always. This is Helmholtz's explanation.

THROUGH TINTED EYEGLASSES

Looking through red-tinted eyeglasses at a *red* inscription on *white* paper you see nothing but a plain red background. The letters disappear entirely from view, merging with the red background. But look through the same red-tinted glasses at *blue* letters on *white* paper and the inscription distinctly appears in *black*—again on a red background. Why black? The explanation is simple. Red glass does not pass blue rays; it is red because it can pass red rays only. Consequently, instead of the blue letters you see the absence of light, or black letters.

The effect produced by what are called colour *anaglyphs*—the same as produced by stereoscopic photographs—is based precisely on this property of tinted glass. The anaglyph is a picture in which the two stereoscopic images for the right and left eye respectively are *superimposed*; the two images are coloured differently—one in blue and the other in red.

The anaglyphs appear as one black but three-dimensional image when viewed through differently-tinted glasses. Through the red glass the right eye sees only the *blue* image—the one intended for the right eye—and sees it, moreover, in black. Meanwhile the left eye sees through the blue glass only the *red* image which is intended for the left eye—again in black. Each eye sees only one image, the one intended for it. This repeats the stereoscope and, consequently, the result is the same—the illusion of depth.

"SHADOW MARVELS"

The "shadow marvels" that were once shown at the cinemas are also based on the above-mentioned principle. Shadows cast by moving figures on the screen appear to the viewer, who is equipped with differ-ently-tinted glasses, as objects in three dimensions. The illusion is achieved by bicoloured stereoscopy. The shadow-casting object is placed between the screen and two adjacent sources of light, red and green. This produces two partially superimposed coloured shadows which are viewed through viewers matching in colour.

The stereoscopic illusion thus produced is most amusing. Things seem to fly right your way; a giant spider creeps towards you; and you involuntarily shudder or cry out. The apparatus required is extreme-ly simple. *Fig. 134* gives the idea. In this diagram G and R stand for

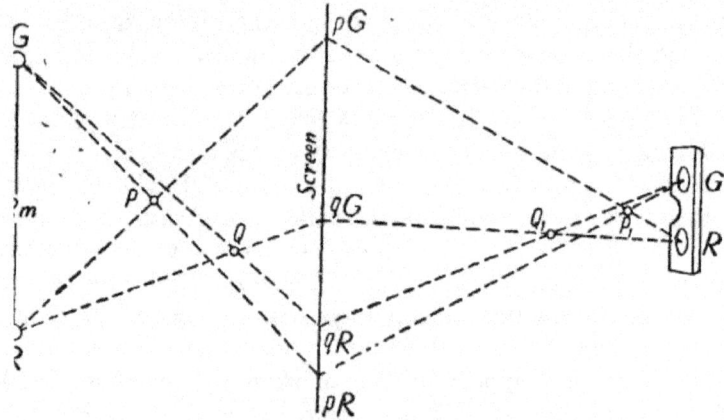

Fig. 134. The "shadow marvel" explained

the green and red lamps (left); P and Q represent the objects placed between the lamps and screen; pG, qG, pR and qR are the tinted shadows that these objects cast on the screen; P_1 and Q_1 show where the viewer looking through the differently-tinted glasses—G is the green glass, and R, the red one—sees these objects. When the "spider" behind the screen is shifted from Q to P the viewer thinks it to be creeping from Q_1 to P_1.

Generally speaking, every time the object behind the screen is moved towards the source of light, thus causing the shadow cast on the screen to grow larger, the viewer thinks the object to be moving from the screen towards him.

Everything the viewer thinks is moving towards him from the screen is actually moving—on the other side of the screen—in the opposite direction—from the screen to the source of light.

MAGIC METAMORPHOSES

I think it would be appropriate at this stage to describe a series of illuminating experiments conducted at the Science for Entertainment Pavilion of a Leningrad recreation park. A corner of the pavilion was furnished as a parlour. Its furniture was covered with dark-orange antimacassars, the table was laid with green baize, on which there stood a decanter full of cranberry juice and a vase with flowers in it, and there was a shelf full of books with coloured inscriptions on their bindings.

The visitors first saw the "parlour" lit by ordinary white electric light. When the ordinary light was turned off and a *red* light switched on in its stead, the orange covers turned pink and the green tablecloth a dark purple; meanwhile the cranberry juice lost its colour and looked like water; the flowers in the vase changed in hue and seemed different; and some inscriptions on the bookbindings vanished without trace. Another flick of the switch and a *green* light went on. The "parlour" was again transformed beyond recognition.

These magic metamorphoses will illustrate Newton's theory of colour, the gist of which is that a surface always possesses the colour of the rays it diffuses, rather than of the rays it absorbs. This is how Newton's compatriot, the celebrated British physicist John Tyndall, formulates the point.

"Permitting a concentrated beam of white light to fall upon fresh leaves in a dark room, the sudden change from green to red, and from red back to green, when the violet glass is alternately introduced and withdrawn, is very surprising ... question of absorption."

Consequently the green tablecloth shows up as green in white light because it diffuses primarily the rays of the green and adjacent spectral bands and absorbs most of all the other rays. If we direct a mixed red and violet light at this green tablecloth, it will diffuse only the violet and absorb most of the red, thus turning purple. This is the main explanation for all the other colour metamorphoses in the "parlour".

But why does the cranberry juice lose all colour when a red light is directed at it? Because the decanter stands on a white runner laid across the green baize. Once we remove the runner the cranberry juice turns red. It loses its colour (in red lighting) only against the background of the runner, which, though it turns red, we ourselves *continue to regard as white*, both by force of habit and due to the contrast it presents to the purple tablecloth. Since the juice has the same colour as the runner, which we imagine to be white, we involuntarily think the juice to be white too. That is why it appears no longer as red juice but as colourless water. You may derive the same impressions by viewing the surroundings through tinted glasses. (See my *Do You Know Your Physics?* for more about this effect.)

HOW TALL IS THIS BOOK?

Ask a friend to show you how high the book he is holding would be from the floor, if he stood it up on one edge. Then check his statement. He is sure to guess wrongly: the book will actually be half as tall. Furthermore, better ask him not to bend down to show how high the book would come up to, but provide the answer in so many words, with you assisting. You can try this with any other familiar object—a table lamp, say, or a hat. However, it should be one you have grown accustomed to seeing at the level of your eyes. The reason why people err is because every object diminishes in size when looked at edgeways.

TOWER CLOCK DIAL

We constantly make the same mistake when we try to estimate the size of objects that are way above our heads, especially tower clocks. Even though we know that these clocks are very large, our estimates of their size are much less than the actual size. *Fig. 135* shows how large the dial of the famous Westminster Tower clock in London looks when brought down to the road below. Ordinary human beings look like midgets next to it. Still it fits the orifice in the clock tower shown in the distance—believe it or not!

Fig. 135. The size of the Westminster Tower clock

BLACK AND WHITE

Look from *afar* at *Fig. 136* and say how many black spots would fit in between the bottom spot and any of the top spots. Four or five? I daresay your answer will be: "Well, there's not enough room for five but there's certainly enough for four."

Believe it or not—you can check it!—there's just enough room for three, no more! This illusion, owing to which dark patches seem smaller than white patches of the same size, is known as "irradiation". This comes from an imperfection of our eye, which, as an optical instrument, does not quite measure up to strict optical requirements. Its refrangible media do not cast on the retina that sharply-etched outline which one gets on the ground-glass screen of a well-focussed camera. Owing to what is called *spherical aberration*, every light patch has a light fringe which enlarges the retinal image. That is why light areas always seem bigger than dark areas of equal size.

187

In his *Theory of Colours* the great poet Goethe—who, though an observant student of nature, was not always a prudent enough physicist—has the following to say about this phenomenon:

"A dark object seems smaller than a light object of the same size. If we look simultaneously at a white spot on a black background and at a black spot of the same diameter but against a white background, the latter will seem about a fifth smaller than the former. If we render the black spot correspondingly larger, the two spots will seem identical. The crescent moon seems part of a circle the diameter of which would be larger than that of the moon's darker portion—which we sometimes see [the ashen light witnessed when the "old moon is in the new moon's arms"—*Y.P.*]. In dark dress we seem slimmer than in clothes of light tones. Light coming over the rim of something seems to make a depression in it. A ruler from behind which we see a candle flame seems to have a notch in it at this point. The rising and setting sun seem to make a depression in the horizon."

Fig. 136. The gap between the bottom spot and each of the top two seems more than the distance between the outer edges of the two top spots. Actually, they are identical

Goethe was right on every point, with the sole exception that a white spot does not always seem larger than a black spot of equal size by one and the same fraction. This depends solely on from how far away you look at the spots. Why? Just move *Fig. 136* still further away. The illusion is still more striking, because the additional fringe we mentioned is always of the same width. Close up, the fringe enlarges the white area by 10%; further away, it takes up from 30 to even 50% of the white area, because the actual image of the spot is already smaller itself. This also explains why we see the circular white spots in *Fig. 137* as hexagons when viewed from two or three steps away. From six to eight steps away this figure will already seem a typical honeycomb.

To say that irradiation is responsible for this illusion is an explanation that has not quite satisfied me, ever since I noticed that *black* dots

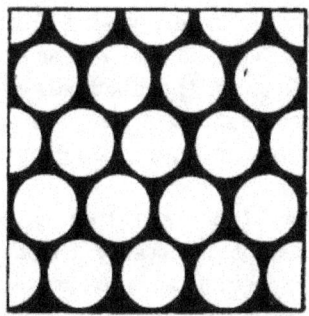

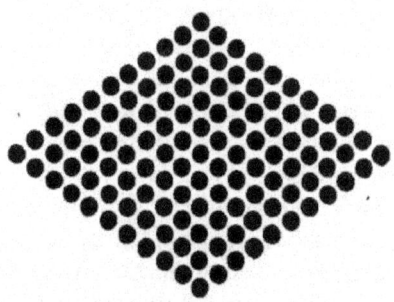

Fig. 137. From a distance the circular white spots seem hexagonal

Fig. 138. From a distance the black dots appear as hexagons

on a white background (*Fig. 138*) also seem hexagonal from far away, though irradiation does not enlarge, but, on the contrary, reduces the dots in size. One must note that the explanations afforded for optical illusions in general are not completely satisfactory. As a matter of fact, most illusions still have to be explained. (For more on the topic see my *Optical Illusions* album.)

WHICH IS BLACKER?

Fig. 139 introduces us to another imperfection of the eye—"astigmatism" this time. Look at it with one eye. Not all four letters will seem identical in blackness. Note which is the blackest and turn the drawing sideways. The letter you thought the blackest will suddenly go grey, and now another letter will seem the blackest. Actually, all

Fig. 139. Look at this word with one eye. One letter will seem blacker than the rest

four letters are identical in blackness; they are merely shaded in different directions. If our eyes were just as perfect and faultless as expensive glass lenses, this would have no effect on the blackness of the letters; but since our eyes do not refract light identically in different directions, we cannot see vertical, horizontal, and slanting lines just as distinctly.

Very seldom is the eye absolutely free of this shortcoming. With some people *astigmatism* is so great that it noticeably lessens the acuteness of vision and they have to wear special glasses to correct this. Our eyes also have other imperfections, which opticians know how to avoid. This is what Helmholtz had to say about them:

"If an optician were to dare sell me an instrument with such imperfections, I would most roundly chide him and demonstratively return the instrument."

Besides these illusions which our eyes succumb to due to certain imperfections in them, there are many other illusions to which they fall victim for totally different reasons.

STARING PORTRAIT

You have most likely seen at one time or another portraits that not only look you square in the eye, but even follow you with their eyes wherever you go. This was noticed long ago and has always baffled many, giving some the jitters. The great Russian writer Nikolai Gogol provides a wonderful description of this in his "Portrait":

"The eyes dug right into him and seemed wanting to watch only him and nothing else. The portrait stared right past everything else, straight at him and into him."

Quite a number of superstitions and legends are associated with this mysterious stare. Actually it is nothing more than an optical illusion. The trick is that on these portraits the pupil is placed square in the middle of the eye—just as we would see it in the eye of anybody looking at us point-blank. When a person looks past us, the pupil and the entire iris are no longer in the centre of the eye; they shift sideways.

190

On the portrait, however, the pupil stays right in the centre of the eye whichever way we step. And since we continue to see the face in the same position in relation to us, we, naturally, think that the man in the portrait has turned his head our way and is watching us. This explains the odd sensation we derive from other such pictures—the horse seems to be charging straight at us however hard we try to dodge it; the man's finger keeps pointing straight at us, and so on and so forth. *Fig. 140* is one such picture. They are often used to advertise or for propaganda purposes.

Fig. 140. The mysterious portrait

MORE OPTICAL ILLUSIONS

There doesn't seem to be anything out of the ordinary in the set of pins in *Fig. 141*, does there? However, lift the book to eye level and, cupping one eye, look at the pins so that your line of vision slides along them, as it were. Your eye must be at the point where the imagined continuations of these pins cross. Then the pins will seem to be stuck in the paper upright. When you shift your head sideways, the pins seem to sway in the same direction.

This illusion is governed by laws of perspective. The drawing is of upright pins projected on paper as they appear to the observer when viewed from the given point.

Our ability to succumb to optical illusions should not at all be regarded as just an imperfection of our eyesight. This ability presents a definite advantage, often overlooked, which is that without it we would have no painting; nor, in general, would we derive any pleasure from the fine arts. Artists draw extensively on these imperfections of our vision.

"The whole art of painting is based on this illusion," the brilliant 18th-century scholar Euler wrote in his famous *Letters on Various*

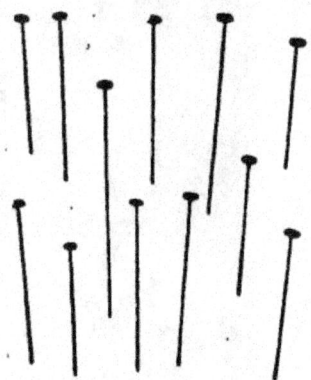

Fig. 141. Fix one eye (have the other shut) at the point where the imagined continuations of the pins would converge. The pins will seem to be stuck in the paper upright. By gently shifting the book from side to side, you get the impression that the pins are swaying

Physical Subjects. "If we passed judgement on things by what they really were, this art (painting) could not exist and we would be blind. The painter would strive in vain to mix his colours, for we would say here is red and there is blue, here is black and there are dashes of white. Everything would be contained in one plane; no difference in distance would be observed and no object could be depicted. Whatever the painter would want to show would all seem to us as writing on paper. And given this perfection, would we not be deserving of pity on being robbed of the delight such pleasant and useful artistry affords us daily?"

There are very many optical illusions, enough (to fill albums (the *Optical Illusions* album mentioned earlier contains more than sixty). Many are common, others are less known. I shall give you some of the more curious instances that are less known. The illusions provided by *Figs. 142* and *143*, with lines on a checkered background, are particularly effective. One simply can't believe that the letters in *Fig. 142* are straight and it is still harder to believe that the circles in *Fig. 143* are not one spiral. The only way to check it is to apply a pencil and trace the circles. Only a pair of compasses will tell us that the straight line *AC* in *Fig. 144* is just as long as *AB* and not shorter, as it appears to be. The other illusions in *Figs. 145, 146, 147*, and *148* are explained in the captions. The following curious incident shows how effective the illusion *Fig. 147* provides is. When the publisher of a previous edition of this book was examining the cliché, he thought it badly done and was about to return it to the printshop to have the grey splotches at the intersection of the white lines scraped off, when I chanced to intervene and explained the matter.

Fig. 142. The letters are upright

Fig. 143. This seems a spiral; actually the curves are circles, which you can see for yourself by following the lines with a pointed pencil

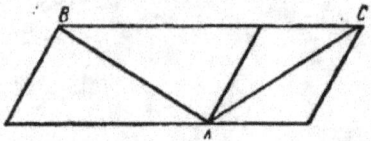

Fig. 144. *AB* is equal to *AC*, though *AB* seems longer

Fig. 147. Tiny faint grey squares seem to appear and disappear where the white strips cross, though the strips are really white throughout, as can be demonstrated by closing up the black squares with a piece of paper. The illusion is due to contrasts

Fig. 145. The slanting line seems broken

Fig. 146. The white and black squares are identical, as, too, are the round white and black spots

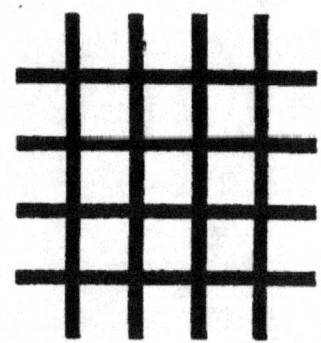

Fig. 148. Faint grey squares seem to appear and disappear where the black strips cross

SHORT-SIGHTED VISION

With his spectacles off, a short-sighted person sees badly. But what he sees and how he sees it is something of which people with normal eyesight have a very hazy notion. Since many are short-sighted, it would not be without interest to learn how they see.

Firstly, to the short-sighted person everything seems blurred. What to a person with normal eyesight are leaves and twigs—all clearly etched against the sky—are to the short-sighted merely an amorphous mass of green. He misses the minor details. Human faces seem younger and more attractive; crow's feet and other minor blemishes are not seen; the coarse ruddiness that may be the product of nature or make-up appears as a delicate flush. He may miscalculate age, being as much as 20 years out. He has odd taste—to those with good eyesight—of the beautiful. He may be considered tactless when he looks a person straight in the eye but seems reluctant to recognise him. He is not to blame. It is his near-sightedness that is the culprit.

"At the Lycée," the 19th-century Russian poet Delvig wrote, "I was forbidden to wear spectacles and my female acquaintances seemed exquisite creatures. How shocked I was after graduation!" When your short-sighted friend (minus his spectacles) chats with you he doesn't see your face, or, at any rate, what you think he sees. His image of you is blurred. No wonder he fails to recognise you an hour later. Most short-sighted people recognise others not so much by their outer appearance as by the sound of their voices. Inadequate vision is compensated for by acuter hearing.

Would you like to know what the near-sighted sees at night? All bright objects—street lanterns, lamps, lighted windows, etc.—assume enormous proportions and transform the world around into a chaotic jumble of shapeless bright splotches and dark and misty silhouettes. Instead of a row of street lamps the short-sighted sees two or three huge bright patches, which blot out the rest of the street. He cannot make out an approaching motor car; instead he sees just the two bright halos of its front lights and a dark mass behind. Even the sky seems different. The short-sighted sees stars of only the first three or four stellar magnitudes. Consequently, instead of several thousand stars he sees only a few hundred, which seem to be as large as lamps. The moon seems tremendous and very close, while a crescent moon takes on a phantastic form.

The fault lies in the structure of the eye; the eye-ball is too deep, so much so that its changed refractive power causes images from distant objects to be focussed before they reach the retina. The blurred retinal image is produced by diverging beams of light.

CHAPTER TEN

SOUND AND HEARING

HUNTING THE ECHO

Mark Twain tells a very funny story of the misadventures of a man whose hobby was to collect—you'll never guess—echoes! This eccentric spared no effort to buy up every tract of land that would have a multiple echo or some other extraordinary natural echo.

"His first purchase was an echo in Georgia that repeated four times; his next was a six-repeater in Maryland; his next was a thirteen-repeater in Maine; his next was a nine-repeater in Kansas; his next was a twelve-repeater in Tennessee, which he got cheap, so to speak, because it was out of repair, a portion of the crag which reflected it having tumbled down. He believed he could repair it at a cost of a few thousand dollars, and, by increasing the elevation with masonry, treble the repeating capacity; but the architect who undertook the job had never built an echo before, and so he utterly spoiled this one. Before he meddled with it, it used to talk back like a mother-in-law, but now it was only fit for the deaf and dumb asylum."

Joking apart, there are some wonderful discrete multiple echoes in various—primarily mountainous—spots, some of which are of long-standing universal fame. The following are some of the better-known echoes. The Woodstock castle echo in England repeats seventeen syllables quite distinctly. The ruins of the Derenburg castle near Halberstadt echoed 27 syllables before one of its walls was blown up. There is one definite place in the rocky cirque near Adersbach in Czechoslovakia which echoes seven syllables thrice; however, a few steps aside even a gunshot will fail to produce any response. There was a castle near Milan that had a very fine repeater-echo before it was demolished.

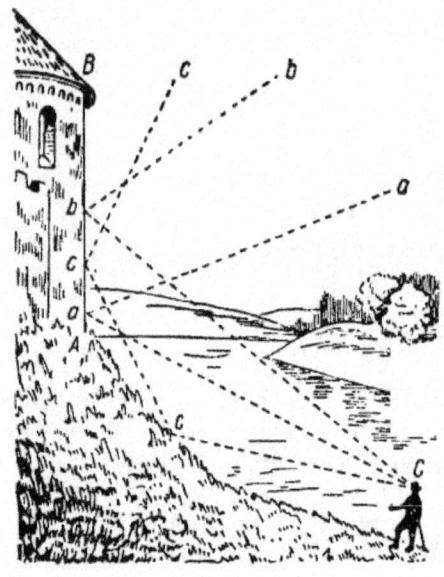

A shot fired from a window in one of its wings echoed back 40 to 50 times, and a word said in a loud voice, some 30 times.

It is not so easy to find even a discrete single echo. The U.S.S.R. is a bit better off in this respect as it has many open plains ringed by woods and many forest clearings, where a shout will already produce a response in the form of a more or less distinct echo coming back from the wall of forest. In mountain land echoes are more varied than in plains, but occur much more seldom and are harder to catch. Why

Fig. 149. There is no echo

is this so? Because an echo is nothing but a train of sound waves reflected back by some obstacle. Sound abides by the same laws as light: its angle of incidence is equal to its angle of reflection.

Imagine yourself at the foot of a hill (*Fig. 149*), with the sound-reflecting barrier AB higher than you are. Naturally the sound waves propagating along the lines Ca, Cb and Cc will not reflect back to your ear but into the air along the directions aa, bb and cc. On the other hand, when the sound-reflecting barrier is at the same level with you or even a bit lower, as in *Fig. 150*, you will hear an echo. The sound travels down along Ca and Cb and returns along the broken lines $CaaC$ or $CbbC$, bouncing off the ground once or twice. The pocket between the points acts like a concave mirror and makes the echo still more distinct. Were the ground between the two points C and B a bulge the echo would be very faint and might not reach you at all, because it would diffuse sound just as a convex mirror diffuses light.

You must develop a certain knack to detect an echo on uneven ter-

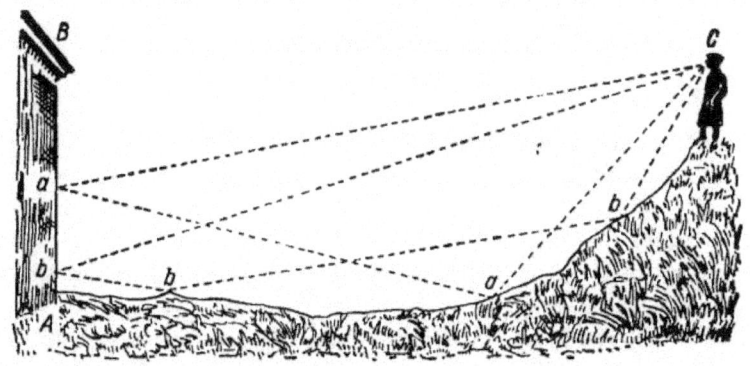

Fig. 150. There is a distinct echo

rain, and even then you must also know how to produce it. In the first place, don't stand too close to the obstacle. The sound waves must travel a long enough distance because otherwise the echo will occur too early and merge with the sound itself. Since sound propagates with the speed of 340 m/sec, at a distance of 85 metres away the echo should be heard exactly half a second later. Though every sound has its echo, not every echo is as distinct, depending on whether it is a beast roaring in a forest, a bugle blowing, thunder reverberating, or a girl singing. The more abrupt and louder the sound, the more distinct is the echo. A hand-clap is best. The human voice is less suitable, especially a man's voice. The high-pitched voices of women and children furnish a more distinct echo.

SOUND AS RULER

Sometimes one can use one's knowledge of the velocity with which sound travels in air to measure the distance to an inaccessible object. Jules Verne provides a case in point in his *Journey to the Centre of Earth*, where in the course of their subterranean exploration the two travellers, the professor and his nephew, lost each other. They hallooed, and when they finally heard each other the following conversation took place between them.

"'Uncle,' I [the nephew] spoke.

"'My boy,' was his ready answer.

"'It is of the utmost consequence that we should know how far we are asunder.'

"'That is not difficult.'

"'You have your chronometer at hand?' I asked.

"'Certainly.'

"'Well, take it into your hand. Pronounce my name, noting exactly the second at which you speak. I will reply as soon as I hear your words—and you will then note exactly the moment at which my reply reaches you.'

"'Very good; and the mean time between my question and your answer will be the time occupied by my voice in reaching you....'

"'Are you ready?'

"'Yes.'

"'Well, I am about to pronounce your name,' said the professor.

"I applied my ear close to the sides of the cavernous gallery, and as soon as the word 'Harry' reached my ear, I turned round and, placing my lips to the wall, repeated the sound.

"'Forty seconds,' said my uncle. 'There has elapsed forty seconds between the two words. The sound, therefore, takes twenty seconds to travel. Now, allowing a thousand and twenty feet for every second, we have twenty thousand four hundred feet—a league and a half and one-eighth.'"

Now you should be able to answer this question: How far away is the train engine if I hear its toot one and a half seconds after I see the wisp of smoke rise from the whistle?

SOUND MIRRORS

A forest wall, high fence, building, mountain, or any echo-producing obstacle in general is nothing but a sound *mirror*, as it reflects sound in the same way as an ordinary flat mirror reflects light.

You can also have a concave sound mirror that would focus the wave-trains of sound. With two soup dishes and a watch you can stage the following illuminating experiment. Put one dish on the table and hold the watch a few centimetres above its bottom. Hold the other dish near your ear as shown in *Fig. 151*. If you gauge the position of all three objects right—do this by trial and error—the ticking of the watch will seem to come from the dish near your ear. But shutting your eyes you enhance the illusion and your ear alone will not tell you in which hand you are holding the watch.

Mediaeval castle-builders often played tricks with sound, by placing a bust either at the focus of a concave sound mirror or at the tail end of a speaking pipe cunningly concealed in the wall. *Fig. 152*, which has been taken from a 16th-century book, shows these arrangements. The vaulted ceiling reflects to the bust's lips all sounds coming in through the speaking pipe; the huge bricked-in speaking

Fig. 151. Concave sound mirrors

pipes carry sounds from the courtyard to the marble busts placed near the walls in one of the galleries, etc. The illusion of whispering or singing busts is thus produced.

Fig. 152. Whispering busts (from a book by Athanasius Kircher. 1560)

SOUND IN THEATRE

The theatre- and concert-goer knows very well that there are halls with good acoustics and bad acoustics. In some speech and music carry distinctly to quite a distance; in others they are muted even quite near.

Not so long ago the good acoustics of one or another theatre was considered simply a stroke of good luck. Now builders have found ways and means of successfully suppressing objectionable reverberation. Though I shall not expand on this point as it can interest only the architect, let me note that the main way of avoiding acoustical defects is to create surfaces to absorb superfluous sounds.

An open window absorbs sound best—just as any aperture is best for absorbing light. Incidentally, a square metre of open window has been accepted as the standard unit to estimate sound absorption. The audience itself is a good sound-absorber, with every person being equivalent to roughly half a square metre of open window. "The audience literally absorbs what the speaker says," one physicist said; it is just as true that the absence of an absorbing audience is literally a great annoyance for a speaker.

When too much sound is absorbed, this is also bad, as, firstly, it mutes speech and music, and, secondly, suppresses reverberation so much that the sounds seem ragged and brittle. As we see, some measure—neither too long, nor too short—of reverberation is desirable. This measure cannot be the same for all halls and must be quantitatively estimated by the designing architect.

There is another place in the theatre of interest from the angle of physics. This is the prompt box. It always has the same shape—have you ever noticed that? Physics is responsible. Its ceiling—a concave sound mirror—serves a dual purpose: firstly, to prevent what the prompter is saying from reaching the audience, and, secondly, to reflect his voice towards the actors on the stage.

SEA-BOTTOM ECHO

Echoes were useless until a method was devised to sound sea and ocean depths with their help. We stumbled upon this invention by accident. When in 1912 the huge ocean liner *Titanic* ran afoul of an iceberg and went down with nearly all its passengers, navigators thought of employing the echo during fogs or at night to detect obstacles in a ship's way. Though this failed to achieve its original purpose, it suggested a fine method, whereby sea depths could be sounded by the echo from the sea-bottom.

Fig. 153 shows you how this is done. By igniting a detonator against the ship's skin near the keel a sharp signal is sent. The sound pierces the water, reaches the sea-bottom and echoes back. This echo, the reflected signal, is recorded by a sensitive device placed against the ship's skin. An accurate timepiece gauges the time interval between the sending of the signal and the reception of the echo. Knowing how fast sound travels in water, we can easily reckon the distance to the reflecting barrier, or, in other words, ascertain the depth.

Echo depth-sounding completely revolutionised sounding practices. To use the old methods one had to stop the ship; and, in general, they were a very tedious affair. The line was payed out very slowly at the rate of 150 metres a minute and it took the same amount of time to rewind it. For instance, it took about 45 minutes to sound a depth of three kilometres. Echo sounding produces

Fig. 153. Echo depth-sounding

the same result in but a few seconds. Furthermore, we don't have to stop the ship to do it and the result is incomparably more accurate, being never more than a quarter of a metre out—provided the time is gauged with an accuracy down to the three-thousandths of a second.

Whereas the exact sounding of great depths is important for oceanography, the quick, reliable and accurate ascertainment of shallow depths is essential for safe steering, especially in offshore waters.

To take soundings today, people employ not ordinary sounds but extremely intensive "ultra-sounds", which we will never hear as their frequency ranges into several million vibrations a second. These sounds are produced by the vibrations of a quartz plate (a piezo-quartz) which is placed in a quickly alternating electric field.

WHY DO BEES BUZZ?

Indeed, why? After all most insects have no special organ for the purpose. The buzzing, which is heard only while the insect is flying, is produced by the flapping of the insect's wings, which vibrate with a rapidity of several hundred times a second. The wings act the role of a vibrating plate, and any plate vibrating with sufficiently great rapidity—more than 16 times a second—produces a tone of a definite pitch.

It is this that reveals to scientists how many times a second an insect moves its wings in flight. To determine the number of times, it is enough to ascertain the pitch of the insect's buzzing, because each tone has its own vibration frequency. With the aid of the slow-motion camera (mentioned in Chapter One) scientists proved that each insect vibrates its wings with practically the same rapidity on every occasion; to regulate its flight it modifies only the "amplitude" of its wing movement and the angle at which the wing is inclined; it increases the number of wing movements per second only in cold weather. That is why the tone of the buzz remains on one level. The ordinary house fly, for instance—its buzz gives the tone F—vibrates its wings 352 times a second. The bumblebee moves its wings 220 times. A honey bee vibrates its wings 440 times a second (tone A) when not burdened with honey, and only 330 times (tone B) when carrying it. Beetles, whose buzzing is lower-pitched, move their wings much less nimbly. Mosquitoes, on the other hand, vibrate their wings between 500 and 600 times a second. Let me note for the sake of comparison that an airplane propeller averages only some 25 revolutions a second.

AUDITORY ILLUSIONS

Once we for some reason imagine the source of a slight noise to be far away, the noise will seem *much louder*. We frequently succumb to these illusions but rarely pay any heed to them. The following curious instance was described by the American scientist William James in his *Psychology*.

"Sitting reading, late one night, I suddenly heard a most formidable noise proceeding from the upper part of the house, which it seemed to fill. It ceased, and in a moment renewed itself. I went into the hall to listen, but it came no more. Resuming my seat in the room, however, there it was again, low, mighty, alarming, like a rising flood or the *avant-courier* of an awful gale. It came from all space. Quite startled, I again went into the hall, but it had already ceased once more. On returning a second time to the room, I discovered that it was nothing but the breathing of a little Scotch terrier which lay asleep on the floor. The noteworthy thing is that as soon as I recognised what it was, I was compelled to think it a different sound, and could not then hear it as I had heard it a moment before." Has anything of the sort ever happened to you? Most likely it has; I, for one, have observed such things more than once.

WHERE'S THE GRASSHOPPER?

We very often err in determining not how far away the sound is, but the direction from which it comes. We can distinguish pretty well by ear whether the shot was fired to the right or left of us (*Fig. 154*), but we are often unable to determine whether it was fired in front of us or behind us (*Fig. 155*). We often hear a shot fired in front of us as one coming from behind. All we can say in such cases is whether it is near or far—depending on how loud the shot is.

Here is a very instructive experiment. Blindfold your friend and seat him in the middle of a room. Ask him to sit still and not turn his head. Then take two coins and click them against each other, standing meanwhile in the imagined vertical plane that passes between your friend's eyes, and ask him to guess where the sound was made. Surprisingly enough, he will point anywhere except at you. But as soon

Fig. 154. Where was the shot fired? On the right or on the left?

as you leave that plane of symmetry which I mentioned, his guessing will be much better, because his ear closest to you will hear the sound a bit earlier and a bit louder.

This experiment, incidentally, explains why it is so difficult to spot a chirring grasshopper. You hear this shrill singing some two steps away on your right. You turn your head but see nothing, and now hear the grasshopper on your left. Again you turn your head, only to hear the singing come from some other spot. The quicker you turn your head, the nimbler our invisible musician seems. Actually, the grasshopper hasn't moved; you've only imagined it to be hopping about. You have fallen victim to an auditory illusion.

Your mistake is that you turn your head so that the grasshopper occupies its symmetrical plane. As you already know, this readily

Fig. 155.
Where was the shot fired? In front? Or behind?

causes you to blunder in determining the direction. So, if you want to find the grasshopper, the cuckoo, or any other similar distant source of sound, turn your head not in the direction from which it comes but away from it, which, incidentally. is exactly what one does when one "pricks up one's ears".

THE TRICKS OUR EARS PLAY

When we nibble at a rusk we hear a noise that is simply deafening. But for some reason our neighbour makes hardly any noise though he is doing the same. How come? The noise we make is one that only we ourselves can hear and it doesn't annoy our neighbours. The point is that like all solid elastic bodies, the bones of our head are very good conductors of sound. The denser the medium through which sound travels, the louder it is. The sound our neighbour makes when nibbling a rusk is a very light one as it travels through air, but this same sound turns into thunder when it reaches the auditory nerve via the solid bones of your head.

Do the following. Grip the strap-ring of your pocket watch between your teeth and stop up your ears. The bones of your head will amplify the ticking so greatly that you seem to hear the pounding of heavy hammers.

The deaf Beethoven, the story goes, could hear a piano being played by placing one end of his walking stick on it and gripping the other end between his teeth. In the same way deaf people can dance to music, provided there is nothing wrong with their internal ear. The music reaches the auditory nerve via the floor and the bones of the head.

Ventriloquism and the "marvels" it works are all based on the peculiar properties of hearing that have just been described.

The illusion that ventriloquism produces depends wholly on our inability to determine both where the voice is coming from and how far away it is. Ordinarily we can do this only approximately. As soon as we find ourselves in unusual circumstances we already make the crudest of blunders in trying to say where the sound comes from. I, too, couldn't rid myself of the illusion when I was listening to a ventriloquist, even though I very well knew what the matter was.

1. How much slower is a snail than you are?
2. How fast do modern aircraft fly?
3. Can you overtake the Sun?
4. How do we get slow-motion films?
5. When do we move round the Sun faster?
6. Why are the upper spokes of a rolling wheel blurred and the lower spokes seen distinctly?
7. Which points in a train going forward move backwards?
8. What is aberration of light?
9. Why do we lean forwards or shove our feet under a chair when we get up?
10. Why does a sailor waddle?
11. What is the difference between running and walking?
12. How should one jump off a moving car? Explain.
13. Baron Munchausen, that famous teller of "tall stories", claimed he had caught flying cannon balls with his hands. Could he have done that?
14. Would you like to have presents tossed at you when you're driving a car?
15. Does a body weigh more or less when falling than when at rest?
16. Must everything fall back to Earth?
17. Is Jules Verne's description of life inside the projectile, that set off for the Moon, right?
18. Can you weigh things right on faulty scales with correct weights, or on a properly calibrated balance, but with wrong weights?
19. Are the bones of our arm advantageous levers?
20. Why doesn't a skier sink into soft snow?
21. Why is it pleasant to loaf in a hammock?
22. How was Paris shelled in the First World War?
23. Why does a kite fly?
24. Does a stone continue accelerating all the time it drops?
25. What is the greatest speed a parachutist making a delayed jump can achieve?
26. Why does a boomerang boomerang?

27. Can we find out whether an egg is boiled without cracking it open?

28. Where is a thing heavier? Closer to the equator or to the poles?

29. When a seed germinates on the rim of a spinning wheel, in which direction does it stem?

30. What is *perpetuum mobile?*

31. Has a "perpetual motion" machine ever been made?

32. Where does a body immersed in a liquid experience the greatest pressure? From the top, the sides, or the bottom?

33. What happens when a small weight suspended on a piece of thread is dipped into a jar of water balanced on a pair of scales?

34. What shape does liquid take when it weighs nothing? Can you prove this experimentally?

35. Why are raindrops round?

36. Is it true that kerosene oozes through glass or metal? Why do people think it does?

37. Can you make a steel needle float?

38. What is flotation?

39. Why does soap wash dirt off?

40. Why does a soap bubble rise? And where does it rise faster—in a cold or warm room?

41. What is thinner? The human hair or the film of a soap bubble? How many times is one thinner than the other?

42. Water gathers under a glass when it's placed, with a burning piece of paper in it, bottom up on a tray of water. Why does this happen?

43. Why does a liquid rise when sipped through a straw?

44. A stick is balanced by weights on a pair of scales. Will the equilibrium be disturbed if the scales are placed under an evacuated bell?

45. What happens to these scales if placed in liquified air?

46. If you lost your weight, but your clothes didn't, would you fly up into the air?

47. What difference is there between a "perpetual motion" machine and a "gift-power" machine? Have any "gift-power" machines been made?

48. What happens to tram rails on a very hot day or on a very cold one? And why is the weather not so dangerous for railway tracks?

49. When do telegraph and telephone wires sag most?

50. What sort of tumblers crack more often because of hot or cold water?

51. Why do lemonade glasses have a thick bottom and why are they no good as tea glasses?

52. What sort of transparent material that wouldn't crack because of heat or cold is best for tableware?

53. Why is it hard to draw a boot on after a hot bath?

54. Can we make a self-winding clock?

55. Can the self-winding principle be used for bigger machines?

56. Why does smoke curl up?

57. What would you do if you wanted to ice a bottle of lemonade?

58. Will ice melt sooner if wrapped in fur?

59. Is it true that the snow warms the Earth?

60. Why doesn't water freeze in underground pipes in winter?

61. Where is it winter in the Northern Hemisphere in July?

62. Why can you boil water in a welded vessel, without fearing that it might come to pieces?

63. Why does a sled cross snow with difficulty in a heavy frost?

64. When can we roll good snowballs?

65. How do icicles form?

66. Why is it warmer at the equator than at the poles?

67. When would we see the Sun rise if light propagated instantaneously?

68. What would happen to telescopes and microscopes if light propagated instantaneously in any medium?

69. Can we make light circumvent obstacles?

70. How is a periscope made?

71. Where should you place a lamp to see yourself better in a mirror?

72. Are you and your reflection in a mirror completely identical?

73. Is the kaleidoscope of any benefit?

74. How can we use ice to light a fire?

75. Can you see mirages in the temperate zones?

76. What is the "green ray"?

77. How should one look at photographs?

78. Why do photographs acquire relief and depth when looked at through a magnifying glass or in a concave mirror?

79. Why is it best to seat oneself in the middle of a movie-house?

80. Why is it better to look at a painting with one eye?

81. How does a stereoscope work?

82. How can we see things like the giants in fairy tales?

83. What is a telestereoscope?

84. Why do things sparkle?

85. Why does the landscape acquire deeper relief when viewed from a passing train?

86. How are stereoscopic photographs of celestial objects taken?

87. What is the effect of the so-called "shadow marvel" based on?

88. What colour does a red flag assume in blue light?

89. What is irradiation and astigmatism?

90. What kind of pictures follow you with their eyes? And why?

91. Is it a person with normal eyesight or a short-sighted person who thinks the bright stars bigger?

92. When you hear the echo 1.5 seconds after you clap your hands, how far away is the sound barrier?

93. Are there such things as sound mirrors?

94. Where does sound propagate faster, in air or in water?

95. To what technical uses can echoes be put?

96. Why does a bee buzz?

97. Why is it so hard to spot a chirring grasshopper?

98. What transmits sound better—air or some denser medium?

99. What is ventriloquism based upon?

* * *

There is a second part to this book. However, both can be read independently of each other.

www.ingramcontent.com/pod-product-compliance
Lightning Source LLC
Chambersburg PA
CBHW071420170526
45165CB00001B/346